TOTAL QUALITY MANAGEMENT IN HIGHER EDUCATION
Symbolism or Substance? A Close Look at the Nigerian University System

Frank Chika Okechukwu, Ph.D.

authorHOUSE®

AuthorHouse™
1663 Liberty Drive
Bloomington, IN 47403
www.authorhouse.com
Phone: 1-800-839-8640

First published by AuthorHouse   2/23/2011

ISBN: 978-1-4520-7646-1 (sc)
ISBN: 978-1-4520-7647-8 (hc)
ISBN: 978-1-4520-7645-4 (e)

Library of Congress Control Number: 2010915083

Printed in the United States of America

This book is printed on acid-free paper.

DEDICATION

This book is dedicated to the ever-green memory of a pacesetter, trailblazer, missionary, and visionary, Professor Innocent Azubuike Okechukwu.

ACKNOWLEDGEMENTS

To God be the glory. Always. My profound gratitude goes to my beautiful and gracious wife, Evangeline, for her support and encouragement. I am very grateful to my children, Chika, Chudi, Chiedu, Chino and Chidera, for putting up with my absences at home while I researched in the library.

My unalloyed gratitude goes to my brother, Joseph Okechukwu, CPA, for his emotional support. For providing me with significant insight into TQM and Nigerian university education, I will remain indebted to Dr. Michael Tegland and Dr. Obinna Ubani.

Finally, this work would not have seen the light of day without the professional publishing support of All Points Graphics in Baltimore.

CONTENTS

Introduction

The genesis of Nigerian university education could be traced to the report of the Elliot Commission on Higher Education in West Africa.[1] The first university college was established in Ibadan, Nigeria, in 1948, during the British colonial regime, with 104 undergraduate students in attendance[2]. The creation of the university college in Ibadan was the springboard for the establishment and growth of other universities in Nigeria. Subsequently, a preponderance of universities have been established by both the federal and state governments of Nigeria to satisfy the manpower and development aspirations of the governments and the desire for university education by the citizens.

The Nigerian constitution requires the federal government of Nigeria to steer "...its policies towards ensuring that there are equal and adequate educational opportunities at all levels."[3] However, with the advent of the military interregnum in the Nigerian body politic, more federal and state universities had been created through the empowering instrument of edicts. All the universities, whether federal or state, were established by federal or state instruments as autonomous institutions.[4] The National Universities Commission (NUC) of Nigeria was established in 1962 to plan strategically and to manage the development of Nigerian universities.

The continuing contribution of Nigerian universities to Nigerian development is today being threatened by four interrelated quality problems. These programs are endemic to both federal and state universities in Nigeria. First, the mix of output of universities is no longer well suited to the requirements for development. Secondly, the quality of those outputs shows signs of having deteriorated. In many instances, the fundamental effectiveness of the outputs may be in doubt. Third, their costs of production are needlessly high (where cost is measured as the other output foregone). Finally, the financing of the outputs is socially inequitable and economically inefficient.[5] Put more dramatically, Nigerian university education "...is in crisis today... the quality of education has dropped as classrooms have become overcrowded and teaching materials increasingly scarce" [6]

Nigerian universities, therefore, need to improve quality, increase efficiency, locate the right output mix, (which could imply smaller enrollment in certain fields of study), and decrease the burden on public sources of financing by encouraging more participation of beneficiaries and their families. Although quality improvements may cost money in the short run, Total Quality Management (TQM) could help save money with time. The Nigerian universities, which gained financially during the years of Nigerian oil boom (1970s and early 1980s), are now confronted with severe economic austerity and the Structural Adjustment Program, which, in turn, adversely affect the quality of Nigerian university education.

Many of the universities in the developed countries that have implemented Total Quality Management find that it not only improved the quality of their graduates, but that it also saved the universities money in the process. Applying TQM to the graduate school admission process at the University of Wisconsin, Madison, Donna E. Shalala, then Chancellor of the university stated:

> In the past, it took an average of 99 days to give the person a yes or no. About 26 of those days were taken up in administrative offices.... We were failing to compete. An improvement team found that most of the waiting involved a lag in getting copies of transcripts from every institution the student had ever

attended. One solution was to redefine what we considered a 'complete file,' and stop waiting for things we didn't need. In administration, we cut the time from 26 to 3 days and saved more than $100,000 in overtime and clerical assistance in one semester. We gave departments the power to admit or reject students within five days of receipt of application. And our students, our customers, got better service.[7]

Chapter 1

The Problem with Nigerian Universities

Nigerian university educators, students, and businesses are becoming more and more dissatisfied. This is because the Nigerian university education system seems inflexible and cannot satisfy the obvious demands of society and the needs of students.

The low quality of Nigerian university education has reached a crisis proportion. The scarcity of funding for capital investment and non-salaried operating expenses has seriously undermined the quality of education in these universities. An example par excellence of this problem is seen at the University of Ibadan.

> For several months now we have been expected to run a physics laboratory without electricity, perform biology and zoology experiments without water and get accurate readings from microscopes blinded by use and age. Chemicals are unimaginably short. The result of all this is a chemistry laboratory that cannot produce distilled water and hundreds of 'science graduates' lacking the benefits of practical demonstrations.[8]

A 1981 report from the Nigerian Commission on salary and Conditions of Service of University staff states that the commission was horrified to witness the disgraceful spectacle of students in the corridors and outside lecture theaters struggling to comprehend the proceedings inside. Fatawo Olatunji commenting on the poor quality of education at the University of Ibadan states:

> Everything in the university today points to an agonizing decline. Students went from their hostels where there are six in a room for two into a dingy lecture room where a teacher shouts his notes across a hall of five hundred listeners"[9]

The Visitation Panel looking into the affairs of Obafemi Awolowo University, Ile-Ife, from 1975 to 1986 reports that:

> Student population at the university as at 1984/85 session was 12,116. Existing hostel accommodation can only take 3,407. The university officially accommodates only 2,766 students. One wonders what happens to the remaining over eight thousand male students.[10]

In a 1995 publication by E.A. Yoloye, the continuation of the problems of Nigerian university education was amplified. Some of the current problems are

(a) an arbitrary increase in the number of universities; (b) inadequate funding of existing ones; (c) pronounced deterioration of existing facilities; (d) massive brain drain of tertiary education teachers; (e) persistent erosion of university autonomy by government; and (f) a serious lowering of the standard of education.[11]

Various reasons have been given for the deteriorating quality of education in Nigerian universities, ranging from misplaced priorities to over-enrollment of students.[12] In particular, faculty brain drain has taken its toll on the university system as many instructors have sought better-paying jobs outside the country. Other quality problems identifiable in Nigerian university education include inadequate classroom space and hostel accommodation, poor strategic financial planning, inadequate student welfare services like medical care and food provision, poor managerial ability on the part of the university management, output of many ill-prepared and deficient students, abandonment of capital and research projects, lack of instructional materials such as books, laboratory equipment and, audiovisual aids, inadequate transportation system for the students, faculty and staff, and accelerated increase in student enrollment without adequate resources.[13]

NIGERIAN UNIVERSITY EDUCATION

There are currently 34 universities, 32 polytechnics, and 55 colleges of education in Nigeria. In 1950, there was only one university college in Nigeria. But the number rose to five in 1962, and thirteen by 1975. The number of universities established within each decade increased. These figures above show an increase of four universities between 1960 and 1970 and 1970 and 1980 and fourteen between 1980 and 1990. This is an indication of a solid average percentage increase of 25 percent per decade.[14]

HISTORICAL DEVELOPMENT OF UNIVERSITY EDUCATION IN NIGERIA:

The purpose of university education in Nigeria has evolved over time and varied from its earliest inception to present.

To address the problem of acute shortage or complete lack of the requisite high-level human resources, Nigeria adopted a two-tier system of approach. The first was to send pre-qualified students abroad to study and train in the universities of Western industrialized countries--- mainly Britain and USA. While this provided an immediate or first-aid solution, subsequently, the second approach was to establish universities and other institutions of higher learning in Nigeria.[15]

The colonial government set up the Nigerian Commission on Post-School Certificate and Higher Education in 1959. The commission was to advise the Nigerian government on the long- term high level human resources needs of the country and to recommend a strategy for university development. This panel, known as the Ashby Commission, was named after its chairman, Sir Eric Ashby. The commission was mandated to conduct an investigation into Nigeria's needs in the fields of post-school certificate and higher education over the next twenty years. The commission was a joint operation of three Nigerians---L.O. Dike, S.D. Onabamiro and Senator Shetima Kashim; three Americans-- R.C. Gustavson, H.W. Hannah, and F. Keppel; and three Britons-- Eric Ashby, J.F. Lockwood, and G.W. Watts. The financing of the commission came from British and American governments, together with the Carnegie Foundation of New York City. The commission recommended the establishment of the National Universities Commission (NUC) to coordinate the development, financing, and other activities of the universities.[16]

The NUC was also responsible for quality control and supervision of Nigerian universities. It set minimum standards for curricula and management of Nigerian universities, reporting directly to the federal Minister of Education. Established in 1962, NUC was also charged with responsibility of effective management and development of the Nigerian university system. Initially, NUC was not a statutory body and was limited merely to advising the government on issues. This accounted for the main reason why most of the recommendations of the commission could not be effectively implemented. When the NUC was converted into a statutory body, it became a powerful agency of the federal government on university education matters.[17]

GROWTH IN ENROLLMENT

University education in Nigeria grew from a low of 338 in 1951 to 63,120 in 1980. Enrollment in the universities grew by 823.4 percent between 1951 and 1961, by 446.5percent between 1961 and 1971 and 1980. During the 1974/5 academic year, the student enrollment in Nigerian universities stood at 26,023 with only six universities in operation. However, by 1977/8 the number of universities had risen to thirteen, with the student population increasing to 48,927. Between 1979 and 1993,twenty-one new universities have been created in the country for a total of thirty four. The year 1975 alone witnessed the establishment of seven new universities in Calabar, Ilorin, Jos, Kano, Maiduguri, Port Harcourt, and Sokoto.[18]

Nearly two thirds of the universities are owned by the federal government while the rest are state-owned. There are no private universities in Nigeria currently. However, private universities did make a brief appearance in 1983 and 1984 when the Supreme Court ruled that there was nothing in the Nigerian constitution and laws of the country preventing the establishing of private universities. Just within six months of the ruling, twenty- six private universities were established or proposed in the country. The federal military government immediately promulgated a decree in 1984 abolishing the existing private universities and prohibiting furtherestablishment of same.[19]

UNIVERSITY STAFFING

The need for reform in Nigerian universities has been felt since the country's independence but particularly since after the Nigerian civil war in 1970. During the National Development Plan Period of 1962-68, most of the universities' academic staff were non-Nigerians. The first decade of Nigeria's independence witnessed 306 out of 516 academic staff as non-Nigerians for the 1962/63 academic session; 422 expatriates out of 670 staff for the 1963-64 session; 503 foreigners out of 1,076 staff for the 1964-65 year; and 595 non-Nigerian out of 1,324 academic staff for the 1966-67 year. This implied 59.3percent expatriate teaching staff in 1962, 63percent in 1963, 46.7percent in 1964, and 44.9percent in 1966. Expatriate staff averaged 53.4percent for the most part of the first decade of independence. How this preponderance of expatriate staff affected the quality of university education at the time was not reported. It suffices to state, however, that the federal government and NUC did not like the status quo and wanted it reformed.

Soon after the civil war, the Nigerian university staffing status changed in favor of the Nigerianization effort via the staff development programs initiated by the universities in the early 1970s. By mid-1977, the expatriate staff constituted only about 26percent of Nigerian universities' teaching staff.

NIGERIA'S EDUCATIONAL GOALS

To better serve the needs of the nation and its citizens, no policy on education can be devised without first identifying the overall philosophy and objectives of the nation. The Federal Republic of Nigeria's National Policy on Education[20] outlined five major national objectives as stated in the second National Development Plan:

1) a free and democratic society;
2) a just and egalitarian society;
3) a united, strong, and self-reliant nation;
4) a great and dynamic economy;
5) a land bright and full of opportunities for all citizens.

The philosophy of Nigeria's education has to agree with Nigeria's national objectives in order to affect national consciousness, national development, national unity, and the acculturation of the individual into a sound and effective citizen. The philosophy is operationally linked to Nigeria's educational objectives:

1) the inculcation of national consciousness and national unity;

2) the inculcation of the right type of values and attitudes for the survival of the individual in Nigerian society;

3) the training of the mind in the understanding of the world around; and

4) the acquisition of appropriate skills, abilities, and competencies, both mental and physical as equipment for the individual to live in and contribute to the development of the society.[21]

These goals form the bedrock of the current national policy on education. They are reflected in the content of the curricula, as well as the planning and administration at all levels. The Nigerian government has a deliberate policy of encouraging graduation of more students in science and technology for industrialization and modernization. The Tannanarive Conference on Higher Education recommended that student enrollment in Science and Arts in African universities should be 60:40.[22]

FINANCING UNIVERSITY EDUCATION IN NIGERIA

The financing of university education in Nigeria is intricately intertwined with the vagaries of the nation's economy. Nigerian economy has undergone some fundamental changes since independence in 1960. Eighty-five percent of Nigeria's foreign exchange earnings came from agriculture. The industrial sector contributed only 3.6percent. However, the discovery and exploitation of oil during the early 1970s bolstered the Nigerian economy. Oil production and sales became the main foreign exchange earner, attracting over 80percent. Subsequently, the agricultural sector was neglected and its contribution to the economy dwindled with impunity. As a result of the oil boom, a period of uncontrolled spending particularly on imported products including food ensued. When the price of oil started to burst in the early 1980s Nigeria's foreign exchange earnings plummeted while foreign debt and foreign debt servicing sky rocketed. Notwithstanding a structural adjustment program with a very stringent fiscal measures adapted by the government in 1986, the national economy has remained in a poor state.[23]

The repercussions on the university have been quite severe: 1) infrastructural facilities have deteriorated and become quite inadequate; 2) gross shortage of instructional materials in institutions including textbooks, journals, chemicals and scientific equipment; 3) decline in the percentage of money allocated to education in yearly federal and state government budgets; 4) the standard of living on the campuses has declined, resulting in overcrowding and poor sanitary conditions; 5) frequent student unrests in demand for better conditions, forcing universities to be closed down to prevent violence; 6) lay off of faculty and staff members because of shortage of funds; 7) imposition of longer hours, larger classes, and heavier workloads on faculty and staff without salary increases or promotions; 8) massive brain drain of staff and students to wealthier countries mostly in the industrialized world.[24]

Chapter 2

Why is Total Quality Management Needed?

To state that the quality of Nigerian university education has suffered greatly on the altar of bureaucracy would not be an understatement. This is, perhaps, why Chris Uroh stated that "the solution to the problem demands a total overhauling of the system, a comprehensive assessment of the problem of individual universities in their separate settings and great steps at solving them."[25]

It would appear, consequently, that a significant Nigerian university improvement is called for at this time, which could involve radical change for it to occur. TQM, by the way, is a systematic process that concentrates on providing the highest value to the customer through the installation of excellence in every aspect of the organization. According to Atkinson, this is done by creating an environment that allows and encourages everyone to contribute to the organization and to develop the skills that will enable them scientifically to study and constantly to improve every process by which work is accomplished.[26]

The general purpose of this book is 1) to identify the key components of TQM for schools, and 2) to assess the implications of schools adopting and utilizing the key components of TQM.

PROMINENCE OF TOTAL QUALITY MANAGEMENT

For some time now, Total Quality Management (TQM), especially as espoused by W. Edwards Deming, has gained prominence in the press, industry, and academic circles. The most important reason for this is not unconnected with the impressive successes now being enjoyed by organizations that have applied the principles of TQM, mainly in the industrialized countries such as Japan, U.S.A., and Canada, etc. These successes were first experienced by industry and later by education. The success of TQM in education has been such that many universities on the cutting edge of societal evolution now offer courses in it while other universities offer degrees in it.

This revolutionary departure from traditional classroom teaching appealed to a number of schools, including the University of Tennessee. Responses from 505 business and engineering schools to a 1992 survey conducted by the Total Quality Forum, an annual conference of corporations and business schools that looks at issues of curriculum, research, and total quality, indicated that about 40 percent have integrated total quality principles into as many as 6 to 10 courses; 45 percent had begun to practice total quality in administrative areas; and 21 percent were actually practicing total quality in the classroom and in research.[27] Kansas Newman College offers eight required courses in TQM in its B.S. program while Marian College, an undergraduate school in Fond du Lac, WI, offers 14 courses in what it calls its quality and productivity management program.[28]

Just how can the theories, principles, and practices of TQM help the sagging quality of education in Nigeria? Is there any need for TQM in Nigeria universities or are there some endogenous or exogenous variables that could effectively mitigate against it. If there are, what are the possibilities of holding those variables constant in order to assess the beliefs of university administrators, faculty members, and students on TQM? How can the teachings of the late W. Edwards Deming assist in continuous improvement of university education in Nigeria? What is the feasibility that an idea that is sweeping developed nations like the United States and Japan can have some positive impact on a developing country like Nigeria as the 21st century draws near? These questions will be addressed in this book. It will not be far fetched to conclude that such a book will be of great significance to policy makers and implementers. It will also help policy monitoring and control processes. Finally, even researchers and scholars might find this pioneering work an invaluable piece of reference.

Chapter 3

An Overview Of Total Quality Management

Available literature on the subject of "Total Quality Management and Nigerian Universities" seems scanty, unorganized, and often contradictory. It is, therefore, the task of this chapter to carry out a scholarly review of some selected literature on this subject.

REVIEW OF LITERATURE RELATING TO THE GENESIS OF TOTAL QUALITY MANAGEMENT (TQM) IN GENERAL

In the glossary of this book, TQM is defined as a system eliciting organization-wide participation in planning and improving processes in order to meet and to exceed customer expectations. It remains to point out that for the purpose of this book, TQM and Total Quality Assurance (TQA) could be used interchangeably.

According to J.M. Juran[29], the genesis of TQM could be traced to the Chinese Shang Dynasty (16th century to 11th century B.C.) when the autocratic imperial family took direct control over the economic functions of the Chinese society. Continuing, Juran (1990) stated that the handicraft industry of ancient China is viewed as being highly structured into broad sectors, then broken down into workshops ranging in sizes from 100 to more than 1,000 craftsmen. Officials were government appointed and responsible for production. Management regulations were strict through to minute division of labor within the workshops, and the results were products that achieve the highest quality standards.

Just as necessity is the mother of invention, the emergence of modern-day TQM resulted from adversity. In 1942, during the Second World War (WWII), the Allied cause suffered. Confronted with unparalleled demand for materiel, the U.S. War Department established a quality control session, staffed essentially by employees from Bell Telephone Laboratories.[30]

However, eleven years earlier, a Bell Laboratories statistician, Walter H. Shewart, had published some interesting ideas on quality control. Aptly observing that all manufacturing processes involved variation, Shewart defined acceptable upper and lower limits for tasks. It was subsequently easy to detect variations outside of these limits and pinpoint their causes. Walter Shewart also introduced the PDCA (Plan-Do-Check-Act) cycle. He then introduced statistical control charts that assisted the employees to plot and to adjust variations.[31]

Shewart's statistics soon made a science out of quality. Inspecting and recording manufacturing according to measurable information brought the process under control and enhanced the productivity of future performance. It also replaced traditional end-line inspection with an on-line awareness of variation.[32]

Additionally, this notion of acceptable quality levels (AQL) obviously appealed to the army procurement officials, who acquired large volumes of armaments from several suppliers. The next logical step was to teach statistical methods quickly to the employees engaged in war time production.[33] Dr. W. Edwards Deming became heavily invested in Shewart's work.

Consequently, by the end of July 1941, Deming had taught many of the total quality courses he would present over the subsequent five years.[34]

Furthermore, an essential recipe in the war effort became recognized as quality techniques that were guarded and protected as military secrets. The files containing these military secrets were declassified and opened to businesses and industries at the end of the war. However, American businesses and industries deluded by the war's legacy of consumer prosperity did not pay a keen enough interest in quality. This, however, proved to be precisely the opposite action with the defeated Japanese.[35]

TQM AND HIGHER EDUCATION

To be relevant to society, universities must grapple with the problems that are significant to society. Today, societies throughout the world are addressing the concept of competitiveness. Universities can contribute to increased competitiveness in education, research, and internal management behavior. TQM is a management philosophy that has spread throughout the world.[36]

Echoing some of the sentiments above, John Hartley wrote that the future of many countries depends heavily upon their institutions of higher education for products such as research and competent intelligence. In the face of growing financial pressure and public criticism, higher institutions must face facts, must lead the effort to change, must take action, and must help formulate new constructive approaches. They must also undergo a program of total quality management to overcome great challenges, he concluded.[37]

At the Fourth Annual Total Quality Forum held in November 1992, five companies and eight universities reported on the progress of the TQM University Challenge. This is a partnership program between business and universities that explores how TQM can be incorporated into university graduate engineering and business curricula. TQM is inherently a cross-functional topic, but faculty and business and engineering schools are specialists. A large portion of the curriculum design and development normally has been done by individual faculty members working alone or in small groups. But the task of building TQM into the curriculum suggests the need for a new approach. Universities should, therefore, design curricula that cover the breath of total quality, integrate total quality throughout the curriculum, and treat TQM as an approach to managing.[38]

According to Martin Di, the TQM concept of continuous improvement (Kaizen) was explored in the University of Herfordshire, using the Juran model of quality improvement as the creation of beneficial change. The practical steps in achieving and maintaining quality improvement in the areas of cash handling, lecture room AV service, and user education were explored. He concluded that TQM can be approached incrementally if there is existing good management practice on which to build its implementation.[39]

A definition of TQM should recognize the fact that it is a continuous process, a development of processes which are error free, and the customers are not simply the "end-users" but also colleagues ("internal" customers). A Quality Improvement Program, based on the principles of TQM, was introduced and operated within the library at the Queen's University of Belfast. The quality improvement program introduced a new method of looking at problem solving, service development and a genuine understanding of the need to satisfy the requirements of all customers.[40]

In a definitional view of TQM that contrasts with the foregoing, J. Wambsganss and Danny Kennett maintained that seeing students as customers is recognized widely but not universally accepted. They supported the view that in the classroom, the student and the instructor are suppliers producing a "product" (knowledge) that a future "customer" (employer or graduate school) will evaluate. They concluded that continuous improvement in an accounting department's curriculum means continually meeting the needs of employer-established quality criteria. Student input should be considered secondary to the needs of the profession.[41]

Pressured by dwindling enrollments and budget worries, a small but growing number of colleges and universities are looking to business and adopting techniques, strategies, and the language of TQM. According to University of Michigan Provost Gilbert R. Whitaker Jr., TQM techniques can help universities use their schedule, facilities, and human resources better.[42]

Although U.S. colleges and universities are recognized worldwide for education and research, a number of problems threaten their strength and stability, Karen Bemowski writes. These include increasing costs and decreasing funding, a decreasing number of high school graduates, and competition. One source of competition is Europe, Japan, and other countries where world- class schools are being built. The second source of competition is major companies such as Motorola and General Electric Co., which are educating their executives internally. Several colleges and universities have recognized their precarious state and have begun using the principles and practices of TQM to improve how they educate and generate knowledge. For example, Columbia University (New York) has incorporated TQM into its curriculum with education modules, courses on TQM, and a TQM master's degree program.[43]

An increasing number of companies are letting business schools know that MBAs who are not trained in total quality management may be passed over for recruitment. Companies want graduates with TQM concepts ingrained. To help achieve that goal, industry is reaching out to schools to drive home the need for TQM in the classroom.[44] At Oregon State University, President John V. Byrne made a personal commitment to lead the university into a TQM program. The implementation of TQM at the university moved through nine phases: 1. Exploration of the TQM concept, 2. The formulation of a TQM pilot study team, 3. The definition of customer needs through quality function deployment, 4. The use of top management breakthrough planning, 5. The use of breakthrough planning by divisions, 6. The formation of daily management teams, 7. The creation of cross-functional pilot projects, 8. The use of cross-functional management, and 9. The introduction of monthly reports, team recognition strategies, and award programs.[45]

Major accomplishments of TQM at Northwest Missouri State University include: the installation of the first comprehensive electronic campus in the U.S., major writing assignments increased 72percent, the semester was lengthened from 15 to 17 weeks, and an assessment program was initiated.[46]

In a year when inflation hovered between two percent and three percent, average annual tuition at public four-year institutions, according to the College Board (a non-profit organization), rose eight percent for the academic year 1993-94 to $2,334 after jumps of 10 percent and 13 percent the previous two years. Tuition for the academic year 2000-01 could reach $3,728 annually for four-year state colleges, $4,624 annually for flagship public universities, and $18,845 for private four-year colleges. Seven in ten institutions report use of TQM principles according to the American Council on Education.[47] Yet only one in ten admits to extensive use of TQM, often focusing on only administrative operations, stopping short of the academic side of the ledger. The time when higher education's noble mission and intellectual value afforded special dispensation is gone; colleges and universities will have to prove their value in the free market. Universities too slow to adapt are as doomed as the horse in the face of the automobile revolution.[48]

QUALITY FROM THE CRITICAL VANTAGE POINT OF THE GURUS

Quality has been defined variously by different experts to suit their temperaments. Philip Crosby, founder and chairman of Philip Crosby Associates, Inc., for example, defined quality as conformance to requirements. This implies knowing what the customer desires, describing that desire, and then precisely meeting that desire.[49]

In a view slightly different from Crosby's, Joseph Juran,[50] founder and chairman emeritus of Juran Institutes Inc., gave two distinctions of quality. In the first instance, quality includes those features of what is being produced that respond to customers' needs and that create the requisite income at the same time. This, according to Juran, is important because, without the requisite income, the exercise would simply be academic.

Also, Juran saw quality as freedom from waste, freedom from trouble, and freedom from failure. The distinction between the two is that higher quality by way of product features generally costs more whereas higher quality in the sense of less failure costs less.[51]

In a view akin to Crosby's and Juran's, Deming[52] notes that quality could have no meaning without some reference to the customer. As a matter of fact, Deming believes strongly that quality is meeting and exceeding the customer's needs and expectations and then continuing to improve.

After WWII, the Japanese were more invested in surviving the disastrous aftermath than in maintaining a high level of their products. Consequently, the quality of Japanese products became so poor and inferior that the term "Made in Japan" symbolized poor, inferior, and shoddy quality.[53]

In sympathy to the defeated Japanese, General Douglas MacArthur recommended that significant changes be made to improve the country's products and their image. He requested assistance from the United States and Dr. Edwards Deming, a government statistician, was sent to teach quality control methods to management leaders in Japan. Deming worked for two years in Japan ---- 1948 to 1950. In recognition of his outstanding and brilliant contributions, the Japanese government honored Deming in 1951 with the creation of the Deming Prize.[54]

Deming's philosophy is based on the concept that everyone should: 1) plan (a production plan is created) 2); do (plan is implemented on a small scale); 3) study (production is studied to ensure that it conforms to plan); and 4) act (lessons learned in the study stage are used to modify the ongoing production process so that a new set of data can be used in creating and implementing the next plan on a larger scale). Then, the cycle must keep rotating. This four-step improvement cycle, which Deming called the "Shewart cycle for learning and improvement," is now commonly referred to as the "Deming Wheel."[55]

Espousing Statistical Process Control (SPC) as a process for monitoring work procedures and setting forth his 14 points for the management of quality and productivity, Deming established quality as goal number one in Japan and set the Japanese 30 years ahead of the United States.[56] Deming's 14 points can be condensed as follows:

1. Create constancy of purpose toward improvement of products and services, with the aim to become competitive and to stay in business and to provide jobs.
2. Adopt the new philosophy. We are in an economic age. Western management must awaken to the challenge, must learn its responsibilities, and must take on leadership for change.
3. Cease dependence on inspection by building quality in the product in the first place.
4. End the practice of awarding business on the basis of price tag. Instead, minimize total cost. Move toward a single supplier for any one item, on a long-term relationship of loyalty and trust.
5. Improve constantly and forever the system of production and service to improve quality and productivity and thus constantly decrease the costs.
6. Institute training on the job.
7. Institute leadership. The aim of supervision should be to help people and machines and gadgets to do a better job. Supervision of management is in need of overhaul as well as supervision of production workers.
8. Drive out fear so that everyone can work effectively for the company.

9. Break down barriers between departments. People must work as a team to foresee problems of production and the use that may be encountered with the product or service.

10. Eliminate slogans, exhortations, and targets for the workplace, asking for zero defects and new levels of productivity.

11. Eliminate work quotas and management by numerical goals. Substitute leadership.

12. Remove barriers that rob the hourly worker of his right to pride of workmanship. The responsibility of supervision must be changed from sheer numbers to quality.

13. Institute a vigorous training of education and self-improvement.

14. Put everybody in the company to work to accomplish the transformation. The transformation is everybody's job.

Deming's 14 points have formed the pillar of TQM because they stress continuous process improvement, on the job training, strong leadership, effective communication, and shared decision making. These elements constitute a program that always benefits the worker as illustrated in the Deming "Chain Reaction": Improve Quality..... Cost Decreases....Productivity Improves....Better Quality....And Lower Prices Capture the Market.....Business Survives And Grows.....More Jobs Created.[29]

Deming's premise is that quality is essential to survival, and he urges manufacturers to work in partnership with their vendors to develop instrumentation and to gain control over their processes. Deming emphasized that the customer is the most important part of the production line.[58]

From 1954 to 1955, another prominent consultant, J.M. Juran, made a series of visits to Japan where he lectured on what is known as Total Quality Control (TQC). According to Juran, quality begins in the design stage and ends only after satisfactory services are provided to the customer. Quality must be viewed as a total all-encompassing concept for a company to be successful, Juran maintained.[59]

Juran attests that his concept should be composed of 90 percent substance and 10 percent exhortation with the formula as follows:

Establish specific goals to be reached.

Establish plans for reaching the goals.

Assign clear responsibility for meeting the goals.

Base the rewards on results achieved.

The "Juran Trilogy" lays out an interrelated strategy for managing quality through Quality Planning, Quality Control, and Quality Improvement. Secondary to the belief that 80percent of problems occur as a result of management inadequacy, Juran insists on training for quality for all managers to enable them to participate in quality improvement projects. The groups are trained in problem solving, brainstorming, group dynamics, and teamwork with a view to determining cause and effect relationships in work-related problems.[60]

Juran's 10 steps to quality improvement are summarized below:

1. Build awareness for the need and opportunity for improvement.
2. Set goals for improvement.
3. Organize to reach the goals.
4. Provide training.
5. Carry out projects to solve problems.
6. Report progress.
7. Give recognition.
8. Communicate results.
9. Keep score.
10. Maintain momentum by making annual improvement part of the regular system and processes of the company.[61]

In his book, *Quality is Free*, Philip Crosby describes quality as free but not a gift. Though not specific on the tools of quality, Crosby works at adjusting people's attitudes toward the definition of quality while maintaining that there must be a commitment of key people to the pursuit of this vital company goal. He also sees managers in the roles of trainers, exemplars, and demonstrators of quality management principles. Training is essential in obtaining commitment, he posits.

Furthermore, Crosby believes in the principle of zero defects. This implies utilizing prevention rather than inspection, testing, and checking. There is no place for statistically acceptable levels of quality in Crosby's concept because this would allow for the belief that errors are planned for or are inevitable. Crosby's 14 steps to quality are summarized below:

1. Management commitment to quality
2. Quality improvement team comprised of department representatives
3. Quality measurement (defect rate)
4. Cost of quality evaluation
5. Quality awareness for all employees
6. Corrective action
7. Ad hoc committee for zero defects program
8. Supervisor training
9. Zero defects day
10. Goal setting
11. Error cause removal
12. Recognition and appreciation
13. Quality councils for regular communication
14. Do it over again to emphasize continuous quality improvement.[62]

All three quality experts (Deming, Juran, and Crosby) see commitment to quality improvement throughout the organization to be the fundamental message. All three also agree on fixing the system rather than the employee, breaking down the work process to remove the barriers to quality, identifying and satisfying customer needs, eliminating waste, instilling pride and teamwork in the organization, and, finally, creating an atmosphere for continuous and permanent quality improvement.

THE JAPANESE MANAGEMENT PHILOSOPHY

The basic influential concepts of Japanese management that affect outcomes in the operation of quality circles have been outlined by several authors. It is a Japanese philosophy that:

1. the group is more important than the individual;
2. workers intelligent enough to do the work are intelligent enough to improve productivity in general;
3. participatory management enhances leadership and motivational skills;
4. all workers form a family unit; and
5. the sharing of feelings in a social atmosphere as opposed to communicating ideas is an important part of a group communication context.[63]

QUALITY CIRCLE DEFINED

Quality circle is a group problem-solving technique in which six to fifteen workers from a given area gather several times a month on company time to study and to solve problems that affect their production. Quality circles use the skills and the know-how of the workers who deal with a problem on a daily basis and whose efforts ultimately determine the quality of the product. Due to the greater potential for worker job satisfaction, the common results from the implementation of quality circles include improved quality of products, lower production costs, better labor/management communication, higher productivity, and increased patents and inventions.[64]

THE QUALITY CIRCLE OBJECTIVES

The Objectives of Quality Circles include:

1. to promote individual job satisfaction,

2. to develop harmonious manager/worker relationships,

3. to improve communications with the organization,

4. to reduce errors and enhance quality of work and product,

5. to create a problem-solving capability within the organization,

6. to increase employee motivation,

7. to promote personal and leadership development,

8. to inspire more effective teamwork,

9. to develop a greater safety awareness on the part of the employees, and

10. to build an attitude of problem prevention.[65]

A difference exists between quality circle and other types of committees and task forces. The specific differences between quality circles and task forces include:

1) Quality circles are voluntary while task forces are usually assigned by management;

2) Development of relationships is an important part of quality circles whereas issues are the focus of task forces;

3) Members work together on a regular basis in quality circles while in a task force members come together for a short time and then disperse;

4) Quality circle activity takes special skills and training while task forces require no special expertise;

5) The work project is developed by members in quality circles, but in task forces, management assigns the work project;

6) Quality circle personnel implement their project whereas a task force may or may not be a part of the implementation process. [66]

PROS AND CONS OF USING CIRCLES

Organizations that become involved in quality circles seem to have many positive outcomes:

1) Quality circles can have a positive effect on the quality of work life.

2) The team approach enhances group spirit and enthusiasm toward reaching a goal.

3) Participatory management is considered by management and participants as a common sense technique.

4) Matching workers' needs to company goals can be accomplished through quality circles,

5) The improvement of quality ensures the improvement of productivity.

6) Recognition of worker participation is a positive reinforcer and a positive motivator.

7) Quality circles have resulted in improved operating effectiveness measured in terms of lowered absenteeism rates, reduced costs, improved product quality, higher morale, and greater job satisfaction.

8) Quality circles can improve productivity and communication.

9) Abraham Maslow's highest hierarchy of needs, that of self-actualization, can be met through quality circles.

10) The design of quality circles provides a vehicle for implementing McGregor's Theory.[67]

On the other hand, many organizations have experienced difficulty in implementing quality circles. Some authors agree that:

1) Some organizations are oriented toward tangible results, and managers are unwilling to allow time for what may seem like intangible service,

2) Lack of objectives can cause lack of direction.

3) Too high of an expectation from the management can cause detrimental pressures.

4) Problems with other unrelated programs can cause problems with circle implementation.

5) Managers often give only "lip service" to implementation of concepts.

6) Realization of goals are inhibited by closed policies.

7) Problems in implementation training could be caused by a poor communication system.

8) Lack of appropriate training causes a breakdown in skill development.

9) Problems could be caused by failure to maintain enthusiasm as well as changes in management.

10) The size of the organization can affect its chances of success or failure.

11) Lack of financial planning can result in loss of funding.

12) Not following up on projects can mean management lack of commitment.

13) Not involving the union can cause problems.

14) Lack of first-hand expertise could cause inadequate expertise.[68]

Blueprint For Success In Quality Circles

Much of the research seems to support the following recommendations for successful circle programs:

1) Organizations must be committed to quality.

2) The focus must be on clear goals and on results of efforts.

3) Responsible facilitator must be available.

4) Advance planning for diffusion and institutionalization is a must.

5) Management on every level must be honestly supportive.

6) Organizations must begin slowly with small pilot programs.

7) There must be a willingness on the part of management to share responsibility.

8) Concerted and comprehensive training of employees to be more effective communicators is essential.

9) Policies and procedures must reflect supportive philosophy.

10) Trust must be the basis of quality circles.

11) Use of organizational development strategies is recommended.

12) Quality circles work best in change-oriented environments.

13) Programs must be on voluntary basis.

14) To ensure access to information and no punishment for errors, open communication must exist.

15) Several needs assessment instruments are recommended to indicate organizational readiness.

16) Employee recognition and feedback are highly recommended.

17) Use of organizational development strategies is a must.[69]

TOTAL QUALITY MANAGEMENT AND EDUCATION

Authors have found remarkable similarities between W. Edwards Deming's philosophies and research from effective schools.

School practitioners and others discovering that total quality processes enhance many of their management strategies, for instance, strategic planning and site based management; TQM is a systematic, all-over approach that provides for the "top down" enablement of "bottom up" decisions and not just another add-on. It empowers employees, organizations, managers, and even whole communities.[70]

It is Rhode's contention that many of the principles of TQM are "naturals" to educators. Embedded in the foundation of TQM are the most primary beliefs about people---how they grow and learn and what motivates them. Educators and others are beginning to realize that TQM makes it flexible and possible for them to change their policies, systems, processes, and practices so as to better achieve their long-held professional and personal values and beliefs. TQM is based on the fact that people are generally already self-improving beings, regularly putting forward their best efforts and need a work setting that supports them.[71]

Chapter 4

Tqm: A Highlight Of The Core Components

The inevitability of the criteria for success in managing the total quality of an organization cannot be over emphasized. Several authors agree on four core components of the TQM processes.[71]

CUSTOMER SATISFACTION

It goes without saying that the customer's role in any organization is very significant.

HERE IS WHAT GLENN HAD TO SAY ON CUSTOMERS:

Customers are worthy people, both honest and competent. It means treating them that way. If our customers are honest and competent people, they are perfectly capable of expressing their valid needs, although we may have to negotiate with them to translate those needs into measurable terms we can work to fulfill. We can ask customers what they want, need, and expect.[72]

Reinforcing his customer argument, Glenn maintained: "All we do is for their sake; without them our work has no purpose. Therefore, if we are serious about quality, customers, no matter whether they are internal or external, have every right to have their requirements, needs, and expectations met the first time and every time."[73] In a view similar to Glenn's, Krone opined that each member of the organization must establish a clear vision of how to provide service to the customer, a vision that must view that service as "courteous, clear, concise, correct, complete and concerned."[74]

The "customer" concept is not new to the professional literature in educational administration. It is more difficult, however, to determine who is the customer in schools than it is in commerce or industry. The list of customers accredited to educators include students, parents, faculty members, board members, employers, community patrons. Students, however, are the primary "customers" of schools. Hence, schools need to meet the holistic needs of their students.[75]

Since students are the school's "customers," they should logically become the focus of the school's "product." The students' need must become the foundation for goal-setting in education. Ultimately, provision of customer satisfaction demands keeping in touch with the customer.[76]

LEADERSHIP

As Axline poignantly pointed out about leadership's role in TQM that when committed leadership is lacking, the various pieces of TQM do not fit together in a coherent pattern.[77] Axline goes on to state that leadership is a prerequisite at all levels of the organization; it should not be confined to or reserved for only the high-ranking executives.

In their contribution, Aalbreqtse and others defined the role of leadership in TQM:

Leadership involves defining the need for change, creating new visions, and using frameworks to mobilize commitment to those visions...frameworks for thinking about strategy, structure, and people. Leadership emphasizes the ability to articulate those visions clearly and forcefully. Leaders provide focus by consolidating or challenging conventional wisdom, and translating their ideas into operational actions.[78]

The role of the leadership is defined in the following terms by Glenn:[79]

1) Leaders excite other people by communication, including action and inspiration.

2) What leaders have in common in addition to their galvanizing vision are positiveness, passion, and humility.

3) Leaders reach beyond mere facts to the what-could-be to facts that have not yet come into existence.

4) When leaders are leading, their focus is outside of themselves, but on the goal...the vision to which they are committed.

Koons further defines the leadership role in TQM, observing that at times it must go well beyond support and facilitation modes:

Not all problems or issues are appropriate for team assignments. There is still a role for creative managers to identify opportunities for operational program enhancements under their control and to take the necessary administrative actions to implement these enhancements. At some points, decisions have to be made even though all of the subordinate staff may not agree. Even in TQM environment, managers are not mere facilitators but still must make some tough decisions that are not always popular.[80]

Commitment has been described as the foundation of an effective TQM initiative. Leadership must play a significant role in promoting commitment. Leaders should, therefore, be charismatic, flexible, and inspiring, especially with those they lead. They should also be able to inspire others to create and to manage change, to take responsibility, and, above all, to take risks. Genuine improvement can be created only when leaders are involving, participating, and actively listening to their followers. Transformational leaders are dependent, are visionary, are inspirational, and are driven by long-term goals, visions, and objectives. They are interested in ends rather than means.[81]

Ultimately, leadership boils down as a function of the leader, the followers, and the situation. The good leader is expected to be able to adapt to the ever-changing circumstances with a view to achieving the organization's set goals and objectives.

PROCESS CONTROL

It goes without saying that quality is the focus of Deming's work:

The central problem in management, leadership, and production is a failure to understand the nature and interpretations of variations. Efforts and methods of improvement of quality and productivity are in most companies and in most government agencies fragmented with no overall competent guidance and no integrated system for continual improvement.[82]

Continuous process improvement shifts the emphasis from problem-solving to investigation of better ways of doing business even when the status quo is acceptable. Continuing in the same line of thought, Scott maintained that the key to TQM is "pursuing a strategy of steady continuous improvement by focusing on and understanding all the elements of existing task. Ideally every person in an organization is always looking for a better way to do a job."[83] He then recommended the use of statistical instruments to minimize variations in processes much as did Deming.[84]

Glasser[85] asserts that although schools have long been characterized by standardized, non-referenced tests to measure quality, these mechanisms do not appear to be compatible with the TQM approach because they generally lack meaningful impact on "production" are imposed by external rather than internal forces, and not to measure quality in ways that are meaningful to the "customer" or student. He concluded that only the student will recognize what represents "quality" (or lack thereof) for him/her and that it is the student's own assessment that should take priority over the assessments of instructors, professors, teachers, parents, administrators, peers, and so on.

STAFF DEVELOPMENT

The importance of staff development certainly has been recognized by TQM. Hunter states: "A final criterion of a profession is that its practitioners never stop learning better ways of providing service for their clients."[86]

Glenn[87] defines these training needs in four basic categories; statistical tools, skills, interpersonal dynamics, and the basic principle of TQM. He believes that all staff development needs must be addressed on an ongoing basis for the whole organizational members in line with the specific roles held and skills required by each of them.

Shanker asserts: 1) Teachers' learning come about through continuous inquiry and interaction with colleagues as well as through exposure to new research ideas from the academic and broader communities; 2) teachers are viewed as an important source of knowledge that should inform what happens in schools; and (3) the school is the focus of staff development.[88]

Team Work

The very essence of TQM is teamwork. This is the ability to work collegially toward a common vision. Management and employees need to trust a common vision. Management and employees need to trust one another and work cooperatively for a common goal. Intra-organizational competition must be discouraged. Teamwork is an integral component for enhanced productivity in an organization. By working together on interdisciplinary and multi-levels, organizations will learn to improve services constantly, to reduce variation in services, and to meet better both the internal and external customer needs.[89]

Reports indicate that people work best when they are in, and feel part of, a team in which they can be trusted and trust each other to do their jobs, share leadership, and make decisions, are accepted and respected, resolve issues with sensitivity and understanding, have the opportunity to accomplish challenging goals, +and contribute to improvement.[90]

Traditional Management and Total Quality Management Compared many literature reviewed here support several main differences between traditional management and total quality management.[91]

IMAGES OF HIGHER EDUCATION IN NIGERIA

American University, Nigeria

University of Nigeria, Nsukka

University of Nigeria, Nsukka

Ogun State University Library

Nigerian Students' Soccer

Improvised Classroom

Flag of the Federal Republic of Nigeria

Official Name	Federal Republic of Nigeria
Capital	Abuja
Population	140 million
Area	923,768 sq km
Currency	Naira ($1=130)
Religion	Islam, Christianity and tribal beliefs
Literacy	57percent
Languages	English, Hausa, Igbo and Yoruba
Major Cities	Abuja, Lagos, Ibadan
Climate	Mainly tropical in nature

The Federal Republic of Nigeria Country Profile
Source: Wikipedia Encyclopedia

Chapter 5

History, Structure, And Administration Of University Education In Nigeria

The demand for university establishment in Nigeria began late in the 19th century when a small group of educated Nigerians campaigned ceaselessly to the British who had conquered and established colonialism over Nigeria to start a university. This group of educated Nigerians consisting of lawyers, doctors, engineers, teachers, and religious leaders felt that if Nigeria were to develop into a modern nation, a university would be indispensable in providing the wide variety of high-level manpower that would be required.

The colonial authorities did not heed to this clarion call until 1948 when the University College, Ibadan (UCI) was established as a college of the University of London. The University College concept implied that UCI could not determine its own examination schemes, could not set or mark its own examination papers, and could not assess candidates for degree awards. Ironically, most of UCI's academic staffers were doctoral products of London University and other renowned British and European universities. For instance, in the 1950/51 academic year,

UCI had thirty-three Ph.D. degree holders: nine from London, three each from Birmingham, Edinburgh, Manchester, Oxford, Reading, Copenhagen, Gottingen, and Toronto universities. Yet the colonial office at that time still did not think the college could design and manage its own academic programs.[92]

During the National Curriculum Reform Conference held in 1969, the former Federal Commissioner for Education Society, was not adequate for Nigeria because it neglected to adapt to the country's cultural and social background. Nnamdi Azikwe had earlier made a similar statement when he argued that Africans under the colonial rule had been "miseducated" to acquiesce in their own subjugation. "The training has so alienated them from their own vital capacities," Azikiwe wrote.[93]

The nationalist struggle in education begun in 1937 by Azikiwe and his American-educated colleagues, and their undaunted spirit gave birth in October 7, 1960 to the University of Nigeria NSUKKA (UNN):

1. An indigenous intellectual center where the raw materials of African humanity will be reshaped into leaders in all fields of human endeavor;
2. A truly African university capable of ridding the renascent Africa of inferiority complex, which has led Nigerians to imitate the excrescences of a civilization that is not rooted in African life;
3. A higher instruction that will be cultural and vocational in its objective and Nigerian in its content.[94]

Following UNN's lead, three other new universities were established in quick succession: The University of Lagos (Unilag) was opened with an enrollment of 102 students. Four months later, the Ahmadu Bellow University (ABU) was opened with 400 students enrolled. The University of Ife was established under the leadership of Chief Obafemi Awolowo in the Western Regions on October 24, 1962 with 244 students enrolled. It was also in 1962 that University College Ibadan became independent of London University and proceeded with its own degree award under the new name of the "University of Ibadan" (UI)[95]

FUNCTIONAL STRUCTURE AND ADMINISTRATION

The primary organizations responsible for the governance of educational institutions in Nigeria are the federal and state Ministries of Education. The Constitution also confers concurrent legislative responsibility for the provision of university, technological, and professional education to the federal and state governments.[96] The Federal Ministry of Education has the responsibility for federally owned educational institutions, and the parastatal organization, NUC, is responsible specifically for the development of federal universities in Nigeria, including some aspects of universities owned by state governments.

The vice-chancellor is the executive head, and the registrar is responsible for the day-to-day administration of universities. The financial affairs of the universities are managed by the bursar's department. The main policy making body of the university is the council consisting of lay and academic persons appointed by the federal government. The university senate is responsible for policies regarding academic matters.

For the purposes of teaching, research, and examination, the universities are grouped into faculties, departments, and committees.

Courses of study leading to bachelor's degrees in Nigerian universities lasts three to four years, depending on the student's entry qualification. Universities offer programs leading to advanced degrees are usually awarded after an additional two years of study. With the master's level of entry qualification, a doctorate degree may be obtained usually after four additional years of study.

Until 1977/8, each Nigerian university had its own admission policy based on such criteria as the quality of entry qualifications and the student's geographical state of origin. There were the problems of duplication of admissions, late acceptances and admissions, and too much competition for admission for particular courses. These and other problems led the federal government to establish a Joint Admission and Matriculation Board (JAMB) to coordinate university admissions and conduct qualifying examinations for university entry.

MAJOR INDICATORS OF EDUCATIONAL QUALITY

Two of the most serious challenges for educational institutions in the 1990s are achieving and maintaining quality, and acquiring, streamlining and maximizing the essential resources to fulfill their respective missions.[97] Most things people do are based on decisions. Decisions, in turn, are based on values, and values provide the ethical, professional, and personal criteria that give direction to thought and deed. The need for constant and never-ending improvement (CANI) necessitates the motivation to question and to evaluate the fundamental values (and assumptions) upon which decisions that directly or indirectly impact the lives of others. It becomes imperative, therefore, that they clearly understand the educational values and underlying assumptions upon which their professional decisions are made and with what results.[98] Nigerian university educators must, therefore, understand the interrelationship between these value decisions and the entire university education system.

When something comes along that suggests that the educational values taught, defended, and practiced for years may be based on faulty assumptions, there is naturally cause for concern, uncertainty, and even strong resistance. Such is the case with the introduction of TQM principles and tenets into education. Universities are reluctant to embrace "efficiency" principles of business into the more "humane" enterprise of education. However, TQM offers much more to the educational scene than simply efficiency measures. It, indeed, offers a whole new way of thinking about education, management styles, and other people including students.[99]

Some of the criteria that could be used to measure the quality of Nigerian university education include but are not limited to:

1. **Staff/Student Ratio:** On the basis of personnel listings of approved positions, the ratio of students to academic staff is 7:1. By way of comparison, the ratio of students to academic staff in Britain is 13:1. One reason for this generous staff comparison with industrialized Britain is the high propensity for Nigerian universities to offer very wide-ranging programs and courses in each institution. Course enrollment of fifteen students is not unusual. Nigerian universities also usually employ large numbers of nonacademic staff, especially to operate municipal and student welfare services and to care for the campus. The University of Nigeria, for example, employs 52,000 staff for a student population of 77,000.[100]

2. **Faculty Stability:** A severe consequence of the economic downturn, and of the concomitant constriction in public budgets, and reduced access to foreign exchange, has been the disappearance from many universities exactly those inputs that make physical plans and highly trained academic staff educationally productive. This has often resulted in the migration of staff and students to other countries especially Europe and America where conditions are deemed much better. Nigerian university administrators should therefore provide "...students with an attractive alternative to (more costly) foreign study, create incentives for university researchers to pursue their work on the continent, and, in so doing, address two aspects of the serious problem of brain drain."[101]

3. **Adequacy of Funding:** The decreasing oil revenue

in Nigeria from the late 1970s through 1980s meant that the government commitment to provide adequate financial resources for universities could no longer be assumed fully. The decrease in Nigerian oil revenue, and Nigerian huge foreign debts together with other fiscal issues led to inadequacy of funding in Nigerian universities, which, in turn, resulted in a lowered quality of education.

In 1978, the report of the Commission of Inquiry into the Nigerian University Crisis found that "the problem of universities hinges on the government trying to develop seven new universities and simultaneously expanding the six older ones."[102] The report of the commission also stated "the commission felt the burden of some aspects of existing policy on the financing of higher education, such as scholarships and bursaries, should be shared between the federal and state governments."[103]

4. **Students Performance:** The most severe outcome of the drying up of non-salary inputs to Nigerian university education is that research ceases and instruction is reduced to little more than rote learning from professorial lectures. This sad situation invariably produces:

> "Chemists who have not done a titration; biologists who have not done a dissection; physicists who have never seen an electrical current; agronomists who have never conducted a field trial of any sort; engineers who have never disassembled the machinery they are called upon to operate; social scientists of all types who have never collected data; lawyers who do not have access to recent judicial opinions; medical students who have never conducted an analysis of their own empirical data; and specialists for whom the programming and use of computers is essential who have never sat before or tested a program; or doctors whose only knowledge of laboratory test procedures is from hearing them described in a lecture hall..."[104]

5. **Students' Accommodations:** It is visibly becoming

increasingly difficult for Nigerian universities to provide enough classroom and hostel accommodations for students. According to the visitation panel looking into the Affairs of Obafemi Awolowo University from 1975 to 1986:

> Student population at the university as of the 1984/85 session was 12,116. Existing hostel accommodation can only take 3,407. The university officially accommodates only 2,766 students. One wonders what happens to the remaining over eight thousand male students.[105]

Furthermore, the panel also reports that students' population has more than doubled over the period under consideration. Consequently, lecture halls can only accommodate a fraction of the students owing to inadequate spaces.[106] This situation invariably adversely affects the quality of this university's education.

6. **University Infrastructure:** Besides inadequate classroom and hostel accommodations, some Nigerian universities also lack modern water facilities. The majority of them have wells around their campuses. However, the water drawn from these wells is not good for human consumption. As a result, many of these students look for water outside the university wells to avoid infectious diseases that can emanate from drinking contaminated well water.

Electricity supply is another problem. This is why students have christened National Electrical Power Authority (NEPA) of Nigeria to mean "Never Expect Power Always." Some universities that have standby generators often suffer from lack of replacement spare parts for the generators when they are broken. This detracts from the quality of Nigerian university education.

Transportation problems are acute in Nigerian universities that operate on multi-campus systems that cannot afford to maintain shuttles or buy new buses for inter- and intra-campus commuting. This results in students scrambling to secure accommodation in the few operational buses, often a spectacle to behold. This has often resulted in accidents and unnecessary loss of students' lives as witnessed by this researcher at Usman Dan Fodiyo, University, Sokoto.

NIGERIA'S ATTEMPT AT QUALITY CONTROL IN THE UNIVERSITIES

Since independence from Britain in 1960, Nigeria has pursued the development and improvement of higher educational institutions including universities. The branch of the government charged with this activity is the Federal Ministry of Education. Among many of its functions, the coordination of educational activities throughout the country is still being pursued

The Federal Ministry of Education has a parastatal that is responsible for advising the federal government of development and improvement of universities in Nigeria. This parastatal is the National Universities Commission of Nigeria (NUC). Through the NUC, the federal government of Nigeria is able to maintain the activities of the Nigerian university system.[107]

The NUC consults with relevant institutions in the preparation of periodic master plans for the balanced and coordinated development of quality universities in Nigeria. It is responsible for the development of federal universities in Nigeria, including some aspects of universities operated by the state governments. NUC is made up of two groups: (1) the commission, which is the highest decision making body of the NUC, made up of members representing various interests within the Nigerian community, and (2) the staff of the direct recruitment or transfer or secondment from public service.

To help improve the quality of university education in Nigeria, the NUC in 1974 was reconstituted by Act No. 1 as a statutory body, giving it the legal authority to perform its functions effectively. In 1979, the Nigerian constitution put higher education on the Concurrent

Legislative List, which is a list of issues on which the state and federal governments can make laws. Eight state governments took advantage of this provision to establish their state universities. In the development of these universities, the NUC played the role of consultant and quality controller.[108] In some state universities, the NUC is on the university council. Most of the universities were established between 1981 and 1983. These universities were established in the states of Anambra, Imo, Rivers, Ondo, Ogun, Lagos, and Cross River. Oyo State University was established in 1989.

At the beginning of the 1980s, about 26 private universities were established in Nigeria. NUC sent out experts to assess the quality of these universities. However, after consultation with the NUC, the federal government abolished all private universities and colleges in 1984 due to the poor quality of the institutions. Decree No. 19 of 1984 prohibited the establishment of private universities in Nigeria. This decree further indicates the efforts of the federal government to provide NUC with increased authority to oversee the emergence of a qualitative university system.

THE POLITICAL CONTEXT IN WHICH THE NIGERIAN UNIVERSITY OPERATES

According to Professor Grace Alele Williams, the political context in which Nigerian universities operate is a complex one. This is because Nigerian universities in all their ramifications are a miniature Nigeria characterized by the internal and external forces of pull and push (positive and negative) reminiscent of a heterogeneous society.[109]

The administration of the university is the joint responsibility of the council and the senate. The council is the highest political body of the university and its membership is inclusive of people from inside and outside the university community. The outsiders who constitute part of the university's external political environment include government appointees representing a variety of interests in the community. The insiders are often in the minority and are members of the university community who gain membership through elections from the senate and congregation. The ultimate financial and managerial responsibilities in the university lie with the council. The senate is the highest academic policy-making body in the university and its main responsibility is to chart the academic programming.

As a member of the council and the senate, the vice-chancellor constantly manages the boundary maintenance problem between the council and the senate. As the head of the academic and administrative staff, the vice-chancellor has to consult with his staff and students regularly, particularly on issues relating to them and report his findings to the council or senate as the case may be. As the only constant variable in university politics, the vice-chancellor is the first citizen of the university community.

The politics associated with the selection of the vice-chancellor could be punctuated by deep rancor, bitter in-fighting, and, at times, open confrontation. Ethnicism and sectionalism could be brought into play depending on the ethnic or ideological composition of the senate. In the final analysis, the appointment of the vice-chancellor is the prerogative of the president of Nigeria after consultation with the Governing Council of the University, and the Federal Ministry of Education. However, after the appointment, the vice-chancellor of the Nigerian university is usually confronted in the beginning with problems arising from the competition for the position. This is because the Nigerian, whether in the academic world, or outside it, is hardly able to take defeat in a sportsman-like manner.[110]

The university is governed by a committee system, which is usually riddled by politics. Provosts, deans, directors, heads of departments, and so on form a group of university administrators upon whom the vice-chancellor must rely for the daily governance of the university. Provosts, deans, and directors serve for two years subject to not more than four years at a time. Heads of departments serve for three years at a time, except the non-professors who serve in acting capacity for not more than one year at a time.

Ethnicism, one of the most common problems of the country, could be found in the university system. For instance, an ethnically conscious registrar could use his position to employ his friends and relatives to most vacancies in the university administration against the rules and regulations governing employment in universities.

Perhaps the most challenging aspect of the internal governance of the Nigerian university is the problem of the student. Generally, students must put up with inadequacies like staff shortage and poor library facilities but hardly with acute shortage of living accommodation like water, food, and so on.[111] This is why a pronounced attention should be given to the social conditions of students along with the advancement of the university. Interactions between the vice-chancellor and his staff and students could reduce potential crisis situations to mere rubbles of arguments.

The different actors in the Nigerian university system, such as the council, senate, academia, students, and so on, all work to maintain university autonomy. Autonomy is very important to the successful fulfillment of the seven principal functions of the university: quality teaching, certification, research, storage of knowledge, publication of texts, public service, and enlightened commentary.[112] However, with pressure from the very top echelon of government, the Nigerian university, which incidentally depends on the government for survival, could hardly sustain its autonomy.

The locality factor in the administration of the university made the Nigerian government to reserve up to 30 percent of the admission vacancies to the locality.[113] Vice-chancellors develop harmonious relationships with the local population through hiring policy as well.

The Nigerian university is called upon to promote national unity. National security itself can be threatened by student unrest that sometimes may arise not from the internal problems of the university but as a response to external political factors. It is, therefore, the hallmark of a good vice-chancellor to understand the political context in which the Nigerian university operates with a view to achieving the university's set objectives while preventing undesired endogenous and exogenous variables.

Deming's 14 point philosophy can be applied to university education in Nigeria in the following way in order to improve the system:

1) **Constancy of purpose:**

W. Edwards Deming's number one point is to create constancy of purpose toward the improvement of products and services by allocating resources for long-term planning, organizational research and workforce education.[114]

Applying Deming's No. 1 point to Nigeria, it can be argued that Nigerian university educators must believe that all resources are aimed at students' development. Therefore, all programs that consume critical resources should be evaluated with a view to eliminating those that do not contribute to student achievement and effectiveness. Faculty, staff, students, support staff, administrators, board members, parents, and the community-at- large must all share a common understanding of desired outcomes, beliefs, and mission as well as a consistent belief that those outcomes can be accomplished. Short-term strategy must be changed to accomplish long-term objectives, and educators must develop a willingness to measure progress.[115]

2) **New Philosophy:**

Deming's No. 2 point is to reject commonly accepted levels of delays, mistakes, defective workmanship and defective materials. Organizations must constantly perfect processes aimed at finding problems, their causes, and ways of correcting them.[116]

This implies a transformation to a new way of thinking and planning for student learning in Nigerian universities. Faculty and staff must reject the idea that students cannot learn at high levels under the right conditions of teaching and learning. NUC and university management must awaken to the challenge, learn their responsibilities, and take on leadership for change.[117]

3) Dependence on mass inspection:

Deming's No. 3 point is to cease mass inspection of purchased materials and services. In Nigerian universities, education cannot wait until the end of the year to measure student progress. The system must understand and use statistical assessment of student growth and place, improve selection processes, and seek statistical evidence of quality[118] development on a daily basis. The emphasis must be a move from the identification of student failure to using continuous improvement to prevent student failure.[119]

4) Awarding business on the basis of price alone:

Deming's No. 4 point suggests ending the practice of awarding business on the basis of price tag.[120] Companies must strive for long-term reduction of total cost rather than piecemeal efficiency.[121]

In this instance, Nigerian university educators should invest in quality, instead of just low cost. With time, high quality produces low cost. Consequently, the universities must choose, use, and evaluate technologies, facilities, new philosophies, textbooks and other resources in instructing based on statistical evidence of success of the particular product and upon accepted outcome criteria.[122]

5) Improve every system:

Deming's No. 5 point suggests searching for problems in the system. Managers are fully responsible for finding and correcting problems in the system.[123]

Nigeria's university educators must acknowledge that improvement is not a one-time event. Commitment to improve the university system constantly necessitates a long-term planning effort. Potential for improvement could be found in each step taken to create or upgrade university programs and services. Nigerian universities should, therefore, consistently identify barriers and seek workable solutions to ameliorate the problems.[124]

6) **On-the-job-training:**

The No. 6 point for Deming recommends on-the-job training with modern training techniques. Unless employees know their jobs and feel free to inform their managers of any problem they encounter, they cannot perform well. Further, statistical methods must be used to identify when on-the-job-training has met its aim.[125]

Nigerian universities need to stay abreast continually of changing demands and requirements. A wide array of external and internal resources must be used for the professional, managerial, and technical development of all university personnel. The limited resources must be channeled toward enhancing student achievement.[126]

7) **Leadership:**

Deming's No. 7 point advocates the institution of leadership. One of the most important responsibilities of managers is supervision. They must learn from employees to help them do a better job.[127]

University educators need to respect the fact that the job of management is not necessarily to tell people what to do but instead to point people in the right direction. They should emphasize the quality of the total program instead of individual behaviors. Evaluations need to be formative, systematic, and programmatic rather than punitive, summative, and personal.[128]

8) **Drive out fear:**

Deming's No. 8 point is to drive out fear so that people can work effectively in an atmosphere of reduced anxiety. Employees must be given job security and encouraged to feel free to ask questions, express ideas, and ask for instructions. An important responsibility of managers is the elimination of fear[129]

This point calls for respect for the basic human dignity of others. One of the best ways to help a student acquire a good self-image is not to do anything to damage it. Drive out fear.[130]

9) **Break down barriers:**

Deming's No. 9 point encourages breaking down barriers between departments. People performing different functions can come together as a team and work effectively to improve products and services.[131]

Universities in Nigeria need to be committed to rebuilding and nurturing an environment in which trust and respect can be applied to what is said, heard, read, and written. Encourage nonthreatening two-way communications on quality outcomes between departments, faculties, and schools. Encourage teamwork for problem solving in order to break down barriers.[132]

10) **Abandon Slogans:**

Deming's No. 10 point advocates the elimination of goals, quotas, posters, and slogans demanding new levels of productivity without the provision of effective methods. Goals must be accompanied by implementational guidelines.[133]

Universities should not want employees searching for excuses and explanations. The employees should rather always strive to improve continually. Slogans asking for perfect performance and new levels of productivity should be eliminated. Most of the causes of low quality and low productivity are inherent in the system and consequently beyond the jurisdiction of university professors and students.[134]

11) **Eliminate numerical goals and quotas:**

Deming's No. 11 point advocates the elimination of work standards that prescribe numerical quotas. Such standards are strongholds against improvement.[135]

Nigerian universities should replace numerical goals with charts that measure progress and do situational analyses. This will be evidence that the school is committed to a long-term process. Mandates and numerical goals should be eliminated, and numbers used constructively. All university employees must be involved in problem identification, program design, planning, budgeting, and materials selection.[136]

12) **Remove barriers that rob pride in workmanship:**

Deming's No. 12 point suggests the removal of barriers that rob employees of their pride of workmanship. Accurate definition of acceptable workmanship is, however, a prerequisite to pride of workmanship. Managers are responsible for definitions.[137]

Nigerian universities need to remove barriers that rob the student, faculty, staff, management, and support staff of their right to pride of workmanship. This includes the abolition of a) annual rating b) merit rating and c) management by objective (MBO). The responsibility of all educated managers must be changed from quantity to quality.[138]

13) **Promote education and self-improvement:**

Deming's No. 13 point advocates the institution of a vigorous program of education and retraining. New jobs and responsibilities will be accorded to people by education and retraining. Improvement in productivity implies reassignment of personnel. Every employee must learn the rudiments for statistical theory and application.[139]

Nigerian universities must provide all employees with training in quality leadership, self-evaluation, measurement, analysis, problem solving, and assertiveness training. Different levels and functions in the organization require different types of training. In-service must, therefore, be a part of the normal work of the university and not just a yearly or monthly event.[140]

14) **Structure management to accomplish the transformation:**

Deming's No. 14 point is to create a structure in top management that will encourage the implementation of the 13 points above on a daily basis.[141]

Nigerian universities leadership must move toward processes that are geared toward problem prevention. To correct deficiencies and to accomplish the complete transformation of the university educational system will take years. Everybody in the system is responsible for assisting to bring about this transformation--students, parents, faculty, staff, board members, teachers, support staff, community partners, administrators, and so on. A thorough comprehension of the past and ability to forecast future needs and requirements all demand an entrepreneurial approach. Conceptual skill is an indispensable recipe in moving from traditional management practices to total quality management practices.

It appears that with the past successes of total quality management in business, industry, and the public sector of the countries where it has been applied, the possibilities of applying TQM to university education in Nigeria does seem to exist. Strategic planning and other school improvement literatures all seem to blend well with the total quality management philosophy.

However, Nigerian universities are users of public finance and consume a substantial portion of the Gross national Product (GNP). The universities constitute a financial burden to government budgets and, currently, governments are finding it hard to provide adequate financial support to operate 34 universities in the country due to the Economic Austerity and Structural Adjustment Program. Reasons ranging from misplaced priorities to over-enrollment of students have been adduced for the deteriorating educational situation in Nigerian universities. How can TQM philosophies hold a promising outcome if successfully applied in Nigerian universities?

Among other things, TQM literature urges that 1) schools need process controls that are valid in order to give feedback for continuous improvement, 2) an individual who is educated in the use of statistical methods and in development of information will be needed to teach faculty, other staff, and administrators how to use information effectively, and 3) that has to be continuous improvement, not just snapshots or random fixes.

In a recent conversation with school administrators before he died on December 20, 1993, Deming maintained:

1. that education can only be transformed one system at a time;
2. that leaders must have a vision and must understand their system in order to put that system into practice;
3. that schools must expect and design for variations among students;
4. that the goal of educational leaders must not be achieving numerical goals but transforming school systems.

The test of anyone's ideas for improving the quality of educational services is whether they can be shown to be effective. Deming and his track record argue persuasively that it is possible to determine whether a system is becoming better or worse, and he provides concepts and tools for sure-footed actions when the latter is the case.

Nigerian university administrators will need to take a serious look at Total Quality Management as one way to bring about needed change and continuous improvement in education as the world gets continuously shrunk into a global village by technology with the concomitant survival of the fittest higher institutions of learning.

End Notes

[1] Keith Hincliff, <u>Higher Education in Sub-Saharan Africa</u> (Bechenham, Kent: Croom Helm Ltd., 1987), p. 110.

[2] Ibid.

[3] <u>The Nigerian Constitution</u>, 1979, Section 19, p.21.

[4] A. Babatunde Fafuknwa, <u>The Growth and Development of Nigerian Universities.</u> Washington, Overseas Liaison Committee, American Council on Education, April 1974, p. 21.

[5] The World Bank, <u>Education in Sub-Saharan Africa: Policies for Adjustment, Revitalization, and Expansion.</u> (Washington, D.C., August 1993), p.70.

[6] Ibid.

[7] Donna E. Shalala, "TQM Application in Education," <u>Executive Excellence,</u> May 1993, p. 6.

[8]Niyi Osundare, "Agonies of a Tottering Tower," West Africa, September 12, 1983, pp. 2121-22.

[9]Ibid. p. 2122.

[10]Federal Republic of Nigeria, Views of the President of Nigeria on the Visitation Panel Report into the Affairs of Obafemi Awolowo University Ile-Ife Oyo State 1975-85. (Lagos: Federal Government Printer, 1989), p. 48.

[11]E.A. Yoloye, "Major Problems for the Year 2000" in the International Encyclopedia of National Systems of Education, 2nd Ed. Edited by T. Neville Postlethwaite (Pergamon, 1995), p. 737.

[12]Ibid.

[13]Dele Omotunde and others, "Education with Tears," Newswatch April 12, 1990, p. 14.

[14]A. O. Sanda, Understanding Higher Educational Administration in Nigeria.

[15]N. Okafor, The Development of Universities in Nigeria, London: Longman, 1971), p. 21.

[16]Commission on Post-School Certificate and Higher Education in Nigeria Investment in Education: Report of the Commission on Post-School Certificate and Higher Education in Nigeria (Ashby Report), 1960, p. 25-30.

[17]Second National Development Plan 1970-74: Programme of Post-War Reconstruction and Development. (Lagos: Federal Ministry of Economic Development and Reconstruction, Central Planning Office), pp. 11-24.

[18]A.M. Ejogu, <u>Landmark in Educational Development in Nigeria</u> (Lagos: Joja Educational Research and Publishers Ltd., Nigeria, 1986), pp. 20-25.

[19]E. Yoloye, "Nigeria: System of Education" in <u>The International Encyclopedia of Education, 2nd Ed.</u> by Torsten Husen and P.N. Postlethwaite (Pergamon, 1994), p. 23.

[20]Second National Development Plan 1970-74: <u>Programme of Post-War Reconstruction and Development</u> (Lagos: Federal Ministry of Economic Development and Reconstruction, Central Planning Office), p. 7.

[21]Ibid., p. 9.

[22]E. Yoloye, "Nigeria: System of Education" in the <u>International Encyclopedia of Education, 2nd Ed.</u> By Torsetn Husen and T.N. Postlewaite (Pergamon, 1994), p. 27.

[23]A. Adedeji, <u>Africa: Autopsy of a Crisis</u>, UNESCO, (May, 1990), p. 9.

[24]K.Twum-Barima, "The University Dilemma," <u>West Africa</u>, September 14, 1987, p. 1786.

[25]Chris Uroh, "Tower Minus Naira = Decay?" <u>Newswatch</u> December 17, 1990, p. 21.

[26]Paul Atkinson, "Leadership: Total Quality and Cultural Change," <u>Management Service</u>, UK, 35 (6), July 1993, pp. 16-19.

[27]Paul Froiland, "TQM invades," Training July 1993, p. 52.

[28]Ibid. p. 54.

[29]J.M. Juran, China's Ancient History of Managing for Quality (New York: McGraw-Hill, 1990), p. 22.

[30]E. Pines, "From top secret to top priority: The Story TQM" Aviation Week and Space Technology, Advertiser sponsored Market Supplement (May 21, 1990), pp. 5, 8, 12, 17, 24.

[31]O. Port, The Quality Imperative (Special Issue), Business Week (October, 1991) p. 15.

[32]Ibid., p. 17.

[33]L. Psihoyos, "The Quality Imperative" (Special Issue), Business Week (October, 1991), p. 20.

[34]Ibid. p. 22.

[35]J. Perline, "America's Quality Coaches" Federal Total Quality Management Handbook (1990), p.8.

[36]R.M. Cybert, "Universities, Competitiveness, and TQM: A plan of action for the year 2000" Public Administration Quarterly Vol: 17, Iss:1, Spring 1993, pp. 10-18.

[37]John Hartley, "Facing the Facts: Reshaping the Academic Enterprise" Vital Speeches 59 (11) March 1993, pp. 337-339.

[38]John J. Kendrick, "Universities, corporations report progress in

integrating total quality into curriculums" <u>Quality</u> 32 (1) January 1993, p. 13.

[39]Di Martin, "Towards Kaizen: The quest for quality improvement" <u>Library Management</u> 14 (4) 1993, p. 4.

[40]Nigel B. Butterwick, "Total Quality Management in the University Library," <u>Library Management</u> 14 (3) 1993, p. 28.

[41]Jacob R. Wambsganss and Danny Kenneth, "Defining the Customer" <u>Management Accounting,</u> May 1995, pp. 39-41.

[42]Gary McWilliams, "The Public Sector: A New Lesson Plan for College," <u>Business Week,</u> Oct. 1991, p. 144.

[43]Kare Bemouski, "Restoring the Pillars of Higher Education," <u>Quality Progress</u> 24 (10) Oct. 1991, p. 37.

[44]Barbara Jorgensen, "Industry to B Schools: Smarten Up on TQM or Else," <u>Electronic Business</u> 18 (13) Oct. 1992, p. 85.

[45]Edwin L. Coate, "TQM at Oregon State University," <u>Journal for Quality and Participation</u> Dec. 1990, pp. 90-91.

[46]Dean L. Hubbard, "Can higher education learn from factories?" <u>Quality Progress</u> 27 (5) May 1994, p. 93.

[47]Ron Gales, " Can Colleges be Re-Engineered?" <u>Across the Board.</u> May 1994, pp. 16, 19.

[48]Ibid., p. 22.

[49]Ibid., 11; Idem, Federal Total Quality Management Handbook.

[50]J. Juran, <u>Quality Planning Analysis</u> (2nd ed.), New York: McGraw Hill, 1980, p. 12.

[51]Ibid., p. 14.

[52]W.E. Deming, <u>Out of Crisis,</u> (Cambridge, MA: Institute of Technology, Center for Advanced Engineering Study, MIT, 1986), p. 21.

[53]R. Dreyfack, <u>Making it in Management the Japanese Way,</u> (Rockville Center, NY: Farnsworth Publishing Company, 1982,) p. 14.

[54]L. Dobyns, "Ed Deming Wants Big Changes, and He Wants Them Fast," <u>Smithsonian,</u> 1990, 10, pp. 74-82.

[55]R.D. Moen, <u>The Deming Philosophy for Improving the Educational Process.</u> Paper presented to the Third Annual International Deming Users' Group Conference, 1989, Cincinnati, OH.

[56]J. Oberle, "Quality Gurus: The Men and Their Message," <u>Training the Magazine of Human Resources Development,</u> January, 1990, pp. 47-52.

[57]Ibid., p. 42.

[58]Dick Schaff, "Beating the drum for quality," <u>Training, the Magazine of Human Resources Development,</u> March, 1991, pp. 5-12.

[59](missing?)

[60]Op. Cit., p. 49.

[61]J.M. Juran, "Quality can't be delegated," <u>Supervision</u> 1988, 49, pp. 6-7.

[62]B. Crosby, "Quality without tears," <u>New America Library,</u> 1984, p. 25.

[63]S. Watanabe, "The Japanese quality control circle: Why it works," <u>International Labor Review,</u> 13 (I), 1991, pp. 57-80.

[64]T.R. Miller, <u>The Quality Circle Phenomenon: A review and appraisal,</u> 54, 1984, pp. 4-7.

[65]C.P. Alexander, "Learning from the Japanese," <u>Personnel Journal</u> 60, 1981, pp. 616-617.

[66]Ibid, pp. 116-17.

[67] L.P. Tang, "The effect of quality circle initiation on motivation to attend quality circle meetings and on task performance" <u>Personnel Psychology,</u> <u>10</u>, 1987, pp. 799-814.

[68]R. Drago, "Quality circle survival: An exploratory analysis," <u>Industry Relation,</u> 27, 1988, pp. 336-351.

[69]R.L. Lansing, The Power of Teams, <u>Supervisory Management,</u> 34, 1989, pp. 39-43.

[70]M. Tribus, "The application of quality management principles in education at Mt. Edgecumbe High School, Stika, Alaska, "<u>An Introduction to Quality Schools. A Collection of Articles on the Concept of Total Quality Management and W. Edwards Deming,</u>" 30, 1990, pp. 1-12.

[71]R.D. Turney, "The application of total quality management to hazard studies and their recording," <u>Internal Journal of Quality and Reliability Management,</u> 8 (6), 1991, pp. 47-53.

[72]T. Glen, "The formula for success in TQM," <u>Bureaucrat</u>, 20, (1), 1991, pp. 19-20.

[73]Ibid., p.17.

[74]B. Krone, "Total Quality Management: An American Odyssey," <u>Bureaucrat</u>, 19 (3), 1990, pp. 35-38.

[75]K.S. Louis and M.B. Miles, <u>Improving the Urban High School: What Works and Why</u> (New York: Teachers College Press, 1990), p. 21.

[76]Ibid., p.93.

[77]L. Axline, "TQM: A Look in the Mirror," <u>Management Review</u>, 80 (7), 1991, p. 64.

[78]R. Aalbregtse, J Hejka, and P. McNeley, "TQM: How do you do it? <u>Automation</u>, 38 (8), 1991, p. 30.

[79]T. Glenn, "The formula for success in TQM," <u>Bureaucrats</u>, 20 (1), 1991, p.18.

[80]P. Koons, "Getting Comfortable with TQM," <u>Bureaucrat</u>, <u>20</u> (2), 119, pp.35-38.

[81]C. Farquhar, "Total Quality Management: A competitive Imperative for the 90's, <u>Optimum</u>, 21(4), 1990, pp. 30-39.

[82]W.E. Deming <u>Out of Crisis</u>, (Cambridge, MA: Institute of Technology, Center of Advanced Engineering Study, MIT, 1986), pp. 465-566.

[83]W. Scott, "TQM expected to boost productivity, ensure survival of U.S. industry," <u>Aviation Week</u>, 13 23), 1989, pp. 64-69.

[84]Ibid., 469; Idem, Out of Crisis.

[85]T. Glenn, "The Formula for Success in TQM," <u>Bureaucrat</u>, 20(1), 1991, p.19.

[86]W. Hoy and C. Miskel, <u>Educational Administration: Theory, Research, and Practice</u> (4th ed.). (New York: McGraw Hill, 1991), p. 26.

[87]F. Jespersen, "Once more with feeling: Quality starts at the top" <u>Business Monthly</u>, 134, 1989, p. 66.

[88]A. O. Sanda, <u>Understanding Higher Educational Administration in Nigeria</u>.

[89]N. Okafor, <u>The Development of Universities in Nigeria</u>, (London: Longman, 1971), p. 21.

[90]W. Hoy and C. Miskel, <u>Educational Administration: Theory, Research and Practice</u> (4th ed.). (New York: McGraw Hill, 1991), p. 26

[91]F. Jespersen, "Once More with Feeling: Quality Starts at the Top," <u>Business Monthly, 134</u>, 1989 p.66.

[92]Ibid., Idem, Landmark in Educational Development in Nigeria.

[93]Nnamdi Azikiwe, <u>Liberia in World Politics</u>, (London: Stockwell), 1934, p. 20.

[94]Nnmadi Azikiwe, <u>Renascent Africa</u>, in Aloy, Ejiogu, <u>Landmarks in Educational Developments in Nigeria</u>, (Lagos Joja Educational Research and Publishers Ltd., 1986), p.77.

[95]Ibid., 79; Idem, Landmark in Educational Development in Nigeria.

[96]Nigeria Constitution, 1979, p. 131

[97]Richard D. DeLosmo, and others, "Rethinking the Connections Between Secondary and Higher Education," Quality and Education: Critical Linkages, Edited by Betty L. McCormick, (Princeton, NJ, 1993), p. 175.

[98]Ibid., p. 145.

[99]Ibid. p. 146.

[100]World Bank, Educational Policies for Sub-Saharan Africa: Adjustment, Revitalization and Expansion. (Washington: World Bank, Educational and Employment Division), Sept. 1987, p. 20.

[101]Ibid., p. 21.

[102]Federal Republic of Nigeria, Government Views on the Commission of Inquiry into the Nigerian University Crisis, 1978. Lagos, Federal Ministry of Information, Printing Division, 1978, p. 4.

[103]Ibid., p. 48.

[104]Ibid., p. 23; Idem, World Bank.

[105]Federal Republic of Nigeria, Views of the President of Nigeria on the Visitation Panel Report into the Affairs of Obafemi Awolowo University, Oyo State 1975-85, (Federal Government Printer, Lagos, 1989), p 48.

[106]Ibid., p. 49.

[107]National Universities Commission, <u>25 Years of Centralized University Education in Nigeria</u>, 1988, p. 97.

[108]Ibid., p. 98.

[109]Professor Grace Alele Williams, "The Politics of Administering a Nigerian University," <u>25 Years of Centralized University Education in Nigeria,</u> 1988, p. 53.

[110]Ibid., p. 55.

[111]Ibid., p. 58.

[112]Ibid.

[113]Ibid.

[114]W.E.. Deming, <u>Out of Crisis</u>, (Cambridge, MA: Institute of Technology, Center for Advanced Engineering Study, MIT, 1986), p. 25.

[115]J. Tilak, "External and Public Investment in Education in Sub-Saharan African," <u>Journal of Education Finance</u>, 15, 1990, p. 470.

[116]Ibid., 30; Idem, Out of Crisis.

[117]J. Leonard "Applying Deming's Principles to our Schools," <u>South Carolina Business</u>, <u>11</u>, (1991), pp. 82-87.

[118]Ibid., 32; Idem, Out of Crisis.

[119]Ibid., 86; Idem, South Carolina Business, 11.

[120]Ibid

[121]Ibid., 33; Idem, Out of Crisis

[122]C. Melvin, Restructuring School by Applying Deming's Management Theories," Journal of Staff Development, 12, 1991, pp. 16-18.

[123]Ibid., 34; Idem, Out of Crisis.

[124]Ibid., 18; Idem, Journal of Staff development, 12.
[125]Ibid., 34 Idem, Out of Crisis.

[126]J. Glaub, "Made in Japan," Illinois School Board Journal, 58, 190, p.5.

[127]Ibid., 35, Idem, Out of Crisis.

[128]Ibid., 6; Idem, Illinois School Board Journal, 58.

[129]Ibid., 35; Idem, Out of Crisis.

[130]W. McLeod, "Toward a system of total quality management (the Deming Way)," An Introduction to Total Quality for Schools. A Collection of Articles on the Concept of Total Quality Management and W. Edwards Deming, 33.

[131]Ibid., 36; Idem, Out of Crisis.

[132]Ibid., 12; Idem, An Introduction to Total Quality for Schools, 33.

[133]Ibid., 33; Idem, Out of Crisis.

[134]Ibid., 8; Idem, Illinois School Board Journal, 58.

[135]Ibid., 8; Idem, Illinois School Board Journal, 58.

[136]Ibid., 36; Idem, Out of Crisis.

[137]Ibid., 12; Idem, Journal of Staff Development, 12.

[138]Ibid., 34, Idem, Out of Crisis.

[139]Ibid., p. 37.

[140] Ibid.

[141]Ibid., 12; Idem, Journal of Staff Development, 12.

Glossary

1. <u>Total Quality Management</u>:   In this context, a structured system for meeting and exceeding needs by creating organization-wide participation in the planning and implementation of continuous improvement processes in accordance with Deming's teachings.

2. <u>Continuous Improvement</u>: Study of processes within an organization to produce constant improvement through evaluation and implementation of ideas, learning, and suggestions, the goal of total quality management.

3. <u>Customer</u>:   Recipient of a product or service from others inside or outside the system.

4. <u>Sub optimization</u>:   Some or all of an organization's processes that impede each other in achieving its stated aim.

5. <u>System</u>:   A network of functions or activities within an organization that work together for a shared goal.

6. <u>University Transformation</u>:   The act of changing the form, outward appearance, condition, nature, or function of universities.

7. <u>Culture</u>:   The concepts, habits, skills, arts, instruments, and institutions of a given people in a given period.

8. <u>Paradigm</u>:   A mindset, pattern, example, or model.

9. <u>Teamwork</u>:   The belief that work can best be accomplished through efforts of more than one person, joint action by a group of people.

10. <u>Quality</u>:   The state of continued excellence.

11. <u>The Presidency</u>:   The Presidency is the Office of the President of Nigeria with its various departments, political, economic, social, or foreign affairs. It also has various other commissions and directorates which have a certain degree of autonomy within its confines. The cabinet office is a branch of the Presidency which coordinates the activities of the services of the Federal Ministry of Education.

12. <u>Federal Ministry of Education</u>: This is the political and administrative agency of the Federal Government of Nigeria responsible for the coordination of the development of education throughout the country.

13. <u>The Minister of Education</u>:   This is the political head and chief accounting officer of the Federal Ministry of Education. The Minister manages the Ministry on behalf of the Nigerian President. The Minister also has

a functional relationship with the NUC and University Administrators.

14. <u>National Universities Commission</u> (NUC):   An organization established in 1962 by the Federal Government of Nigeria under the leadership of Alhaji Abubakar Tafawa Balewa to oversee the activities of the Nigerian university system. This organization was recommended by the Ashby Commission on Higher Education. Initially, the role of NUC was advisory in addition to liaising among universities, government departments and ministries. It is also the channel through which the universities get their subvention from the Federal Government of Nigeria. Government policy matters are channeled to the universities through the NUC. The NUC has a working relationship with both the federal and state Ministries of Education, and the university administrators.

15. <u>University Administrators</u>:   These are the principal officers of the Nigerian university system. They are the university non-teaching staff who were asked their perceptions on Total Quality Management in selected Nigerian universities. They have a functional relationship with the NUC and the Ministry of Education.

i) Vice Chancellor (V.C.) This is the chief administrator and academic officer, the Chief Executive of the university. He is also the chairperson of convocation, congregation, and senate. He is appointed or removed from office by the President of Nigeria after consultation with the university council, (the governing body of the university headed by the chancellor).

ii) Deputy Pro-Vice Chancellor   He acts in the place of the V.C.

when the office of the V.C. is vacant. He is appointed by the council after consideration and recommendation.

iii) Registrar   This is an administrative officer of the university and is responsible to the V.C. for the day-to-day administration of the university. He is the secretary to the council, senate, congregation and convocation. As a secretary for these bodies, he is responsible for the executive action arising from all decisions taken.

iv) Bursar This is the chief financial officer of the university. He is responsible to the V.C. for the day-to-day administration and control of the financial affairs of the university.

v) Academic Planner   This officer works with the V.C. in the academic planning of the institution.

16. <u>Faculty Members</u>:   This is the teaching staff of universities. The dean of the faculty is a professor elected by the faculty board to oversee the academic and administrative affairs of the faculty. The dean presents to convocation qualified persons for the conferring of degrees.

APPENDIX I

List of Higher Institutions in Nigeria

FEDERAL UNIVERSITIES

1.	Bayero University, Kano	P.M.B. 3011, Kano	064-626021
2.	University of Port-Harcourt	P.M.B. 5323, P/H	088-228218/226400
3.	University of Nigeria, Nsukka	Nsukka, Anambra State	042-771911/52
4.	University of Ilorin	P.M.B. 1515, Ilorin, Kwara State	031-221727
5.	University of Ibadan	Ibadan, Oyo State	022-400550
6.	Ahmadu Bello University, Zaria	P.M.B. 1013, Zaria, Kaduna State	069-50581-4
7.	University of Jos	P.M.B. 2084, Jos, Plateau State	073-35951
8.	Obafemi Awolowo University	Ife-Ife, Oyo State	036-230290-99
9.	University of Benin	P.M.B. 1154, Benin	052-200482
10.	University of Lagos	P.M.B. 12003, Lagos	01-821945

11.	Usman Dan Fodio University, Sokoto	P.M.B. 2346, Sokoto	060-232134/236688/232366
12.	University of Maiduguri	P.M.B. 1069, Maiduguri	076-232577
13.	University of Calabar	P.M.B. 1115, Calabar	087-222790
14.	Fed. Uni. of Technology, Akure	P.M.B. 704, Akure	034-233113
15.	Fed. Uni. of Technology, Minna	P.M.B. 656, Minna	066-222397
16.	Fed. Uni. of Technology, Owerri	P.M.B. 1526, Owerri	083-233228
17.	Fed. Uni. of Technology, Yola	P.M.B. 2076, Yola	---
18.	A.T. Balewa University, Bauchi	P.M.B. 248 Bauchi	---
19.	University of Agriculture, Abeokuta	P.M.B. 2240	039-230768/200170-177
20.	University of Agriculture, Makurdi	P.M.B. 2373, Makurdi	044-33204/33205
21.	University of Abuja	P.M.B. 117, Abuja	044-33485
22.	Nigerian Defense Academy	P.M.B. 2109, Kaduna	---

Appendix II

State Universities

1. Anambra University of Technology

2. Bendel State University

3. Cross River State University

4. Imo State University

5. Lagos State University

6. Ogun State University

7. Ondo State University

8. Rivers State University

9. Oyo State University

SELECTED BIBLIOGRAPHY

Adedeji, A. <u>Africa: Autopsy of a Crisis</u>. UNESCO Publication, 1990.

Alexander, C.P. "A hidden benefit of quality circles". <u>Personnel,</u> 63, 1984.

Alikhan, H. Economic Modeling of Structure Adjustment Programs: Impact on Human Conditions. <u>Africa Today,</u> 1990.

Allie, R.E. "The Middle Management Factor in Quality Circle Programs," <u>Advantage Management</u>, 51.

Aquifer, N. <u>The Development of Universities in Nigeria</u> (London: Longman), 1971.

Ashar, H. & Shapiro, J.Z. "Are Retrenchment Decisions Rational?," <u>Journal of Higher Education</u>, 61, 2, 1990.

Askin, S. & Baker, S. "Rethinking Higher Education" <u>African News,</u> 28, 1987.

Atkinson, P. "Leadership: Total quality and cultural change," Management Services (UK), 35 (6), 1991.

Aulbregtse, R.L., J.A. Hejka, and P.K. McNeely "TQM: How do you do it?," Automation, 38, 1991.

Axline, L.L. "TQM : A look in the mirror," Management Review, 80, 64. 1991.

Azam, J.P. et al The Impact of Macroeconomic Policies on the Rural Poor. (New York: Policy Division, United Nations Development Program), 1989.

Baker, J.T. "We're lost, but we're making great time." Industrial Management, 36 (6), 1991.

Berman, S.J. and S.A Hellweg. "Perceived supervisor communication competence and supervisor satisfaction as a function of quality circle participation." Business Communication, 26, 1989.

Berry, J. "Employee morale an added benefit of quality circles," Medical Care, 27 (8), 1984.

Blue, R.R. "Overcoming the threat of participation," Supervision, 50, 1988.Bowman, J.S. "Quality circles: Promise problems and prospects in Florida." Public Personnel Management, 180, 1989.

Chapelier, G and H. Tabatabai. Development and Adjustment: Stabilization, Structural Adjustment and UNDP Policy. (New York: Policy Division, United Nations Development Program), 1989.

Commission on Post-School Certificate and Higher Education in Nigeria. Investment in Education: Report of the Commission on Post-

School Certificate and Higher Education in Nigeria (Ashby Report), 1960.

Dollar, W.E. "The danger of quality circles." Purchasing, 94, 1983.

Dreyfus, J. "Victories in the quality crusade." Fortune, 118, 1988.

Ejiogu, A.M. Landmark in Educational Development in Nigeria Lagos: (Joja Educational Research and Publishers Ltd., Nigeria, 1986).

Fafunwa, A.B. History of Education in Nigeria. (London: George Allen and Unwin Publishers Ltd), 1974.

Farquhar, C.R. "Total quality management: A competitive imperative for the 90s." Optimum, 20 (4), 1990.

Feigenbaum, A.V. Total Quality Control. (New York: McGraw Hill), 1954.

Fuller, B and S. P. Heynman. "Third World School Quality, Current Collapse, Future Potential." Educational Researcher, 1989.

Gabor, A. "Deming's Quality Manifesto." Best of Business Quarterly, 1990.

Gail, C. "The quality imperative" (special issue) Business Week, 1991.

Gist, M., E. Lock, and M. Taylor. "Organizational behavior: Group structure, process, and Effectiveness." Journal of Management, 13, 1987.

Glasser, W. The Quality School. (New York: Harper & Row), 1990.

Glenn, T. "The Formula for Success in TQM." <u>Bureaucrat, 20</u>, 1991.

Griffin, R.W. "Consequences of quality circles in an industrial setting: A longitudinal Assessment." <u>Academic Management</u>, 31, 1988.

Guaspari, J. "You want buy-in to quality? Then you've got to sell it." <u>Management Review</u>, 77, 1988.

Gyrna, F.M. <u>Juran's Quality Control Handbook</u>. (New York: McGraw-Hill), 1988.

Hanson, W.L. and J. Stampe. "The Financial Squeeze on Higher Education Institutions and Students: Balancing Quality and Access in the Financing of Higher Education." <u>Journal of Education Finance</u>, 5, 1989.

Harrington, A.M. "The Cat-22 of Total Quality Management." <u>Across the Board</u>, 28 (9) 1991.

Hinchcliff, K. <u>Higher Education in Sub-Saharan Africa</u>. (Beckenham, Kent: Croom Helm Ltd.), 1987.

Houihan, T.R. "Planning a total quality school district in Johnston County, North Carolina." <u>An Introduction to Total Quality for Schools. A Collection of Articles on the Concepts of Total Quality Management and W. Edwards Deming</u>, 32, 1991.

Hoy, W.K. and C. G. Miskel. <u>Educational Administration Theory, research, and practice (4th ed.)</u> (New York: McCraw-Hill), 1991.

Hughes, M.G. and E.O. Fagbamiye. "Education: The Contrast Between

1973 and 1983." <u>A Review In Contemporary Analysis In Education Series (eds.)</u> (London: The Palmer Press) 1986

Ishikawa, K. and D. Lu. <u>What is total quality control? The Japanese Way.</u> (Englewood Cliffs, New Jersey: Prentice-Hall, Inc.), 1985.

Jesperson, F.F. "Once more with feeling: Quality starts at the top." <u>Business Monthly</u>, 134.
1989.

Johnson, J.G. "Why aren't our managers better leaders?" <u>Tapping the Network Journal</u>, 2 (1) , 1991.

Kizerbo, J, <u>Caught in a trap, the African Educational System.</u> (UNESCO Sources), 1990.

Klekamp, R.C. "Commitment to quality is not enough." <u>Advantage Management</u>, 54, 1989.

Koons, P.F. "Getting Comfortable with TQM." <u>Bureaucrat, 20</u> (2), 1991.

Krone, B. "Total quality Management: An American Odyssey." <u>Bureaucrat</u>, 19 (3) 1990.

Lamb, D <u>The Africans</u> (New York: Vintage Books), 1985.

Lansing, R.L. "The Power of Teams." <u>Supervisory Management</u>, 34, 1989.

Larson, J.S. "Employee participation in federal government." <u>Public Personnel Management</u>, <u>18</u> 1989.

Lawler, E.E. "Quality circles: After the honeymoon." <u>Organizational Dynamics</u>, 15, 1987.

Leader, C. "Making total quality management work: Lessons from industry." <u>Aviation Week and Space Technology</u>, 131 (18), 1989.

Leonard, J.F. "Applying Deming's principles to our schools." <u>South Carolina Business, 11</u>, 1991.

Louis, K.S. and M. B. Miles. <u>Improving the Urban High School</u>: <u>What Works and Why</u>. (New York: Teachers College Press.), 1990

Louri, S. "The Impact of Recession and Adjustment on Education." <u>A Review in Human Development, Adjustment and Growth, (eds.)</u> 1986.

Maguire, M. "Fathoming Deming's Ideas on Quality." <u>The Washington Times</u>, March 1991.

Marx, G. "Total quality management. A new lens to focus schools." <u>Leadership News, AASA, 83,</u> 1991.

McLeod , W. "Toward a system of total quality management to hazard studies and their recording." <u>International Journal of Quality and Reliability Management</u>, 8 (6), 1991.

Meaney, D. "Quest for quality." <u>California Technology Project Quarterly, 2</u>, 1991.

Melvin, C. "Reconstructing schools by applying Deming's Management theories." <u>Journal of staff Development</u>, 12, 1991.

Mendenhall, S. "How the supplier can contribute to TQM." Health Industry Today, 45, 1991.

Mills, C.A. "Austerity Triggers Riots In Nigeria." Africa Report, July 1989.

Moen, R. & Nolan, T. Process improvement. Quality Progress 20, 1987.

National Universities Commission 25 Years of Centralized University Education In Nigeria. (Lagos: Newswatch Communications.), 1988.

National Universities Commission. 20 Years of University Education in Nigeria. (Lagos: National Universities Commission), 1983.

Nemote, M. and D. Lu. Total Quality Control for Management: Strategies and Techniques from Toyota and Toyota Gosei, (Englewood Cliff, New Jersey: Prentice-   Hall, Inc.), 1987.

Nigeria. First National Development Alan, 1962-68: (Lagos Federal Ministry of Economic Development and Reconstruction, Central Planning Office, 1962).

Ogundimu, B.A. "Nigeria" in T.N. Postlewaithe, Eds., The Encyclopedia of Comparative Education and National Systems of Education: (New York: Pergamon Press), 1988.

Perlman, l. "Control data: Giving employees ownership in Quality." Management Review, 78, 1989.

Pines, E. "From top secret to top priority: The story of TQM." Aviation Week and Space
Technology, Advertiser Sponsored Market Supplement, May, 1990.

Port, O. "The quality imperative (Special issue)." <u>Business Week</u>, October 1991.

Psihoyos, L. "The quality imperative (special issue)." <u>Business Week</u>, October 1991.

Rensis Likert Associates. <u>Leaders in Organizational Diagnosis and Human Resource Development</u>. (Ann Arbor: University of Michigan, Institute for Social Research.), 1984.

Rhodes, L. "Toward a system of total quality management (the Deming Way)." <u>An Introduction to Total Quality for Schools. A Collection of Articles on the Concepts of the Total Quality Management on W. Edwards Deming</u>, (Arlington Va.), 1991.

Ritter, D. "Let's elevate quality on our national agenda." <u>National Productivity Review</u>, 10, 1991.

RLA (Rensis Likert Associates) <u>Survey of Organizations-2000: Supervisors Handbook.</u>
(Michigan: Author), 1987.

Sanda, A.O. <u>Understanding Higher Educational Administration in Nigeria</u>, 1991.

Sayegh, R. <u>International Yearbook of Education</u>. Vol. XLI. (France: International Bureau of Education, UNESCO), 1989.

Scherkenbach, W. <u>The Deming route to quality and productivity: Maps and roadblocks</u>. (Washington, D.C. Ceep Press), 1988

Scott, W.B. "TQM expected to boost productivity, ensure survival of U.S. industry." Aviation Week , 13, 1989.

Shanker, A. "Staff development and the restructured school." In B. Joyce (Ed.) Changing culture through staff development (pp. 91-103) (Alexandria, VA: Association for Supervision and Curriculum Development.), 1990.

Smith, P. "Nigeria, Adjustments New Phase." Africa Recovery, 4 (New York: United Nations),
April - June, 1990.

Steimer, T.E. "Activity based accounting for total quality." Management Accounting, 72, 1990.

Tang, L.P. "The effect of quality circle initiation on motivation to attend circle meetings and on task performance." Personnel Psychology, 40, 1987.

Tilak, J.B.G. "External and Public Investment in Education in Sub-Saharan Africa." Journal of Education Finance, 15, 1990.

Tobin, L.M. "The new quality landscape: Total Quality Management." Journal of Systems Management, 41 (11), 1990.

Tribus, M. "The application of quality management principles in education at Mt. Edgecumbe High School, Sitka Alaska." An Introduction to Quality Schools. A Collection of Articles on the Concepts of Total Quality Management and W. Edwards Deming, 23, 1990.

Twum-Barima, K. "The University Dilemma" West Africa,. 1786, Sept. 1987.

UNDP. Education and Training in the 1990's (New York: United Nations Development Program), 1989.

UNESCO Planning and Management of Education at Development: Report of the International Congress. (Mexico City), March 1990.

Uroh, C., "Tower Minus Naira = Decay" Newswatch, Dec. 1990.

Vansina, L.S. "Total quality control: An overall organizational improvement strategy." National Productivity Review, 9, 1989.

Vieta, K.T. "The Agitated Front." West Africa, Jan 16-21, 1989.

Vogt, J.F. and B. D. Hunt. "What really goes wrong with participative work groups?" Training Developments 42, 1988.

Walton, M. The Deming Management Method (New York: The Putnam Publishing Group), 1988.

Wheeler, J. "Financing Human Resource Programming." A Review in Managing Human Development, (eds.), Sept. 1987.

World Bank Educational Policies for Sub-Saharan Africa: Adjustment, Revitalization and Expansion (Washington: World Bank, Educational and Employment Division), Sept. 1987.

World Bank. "Investing in People Is Key to Future." Africa Update. (Washington, D.C.), 1990/91. World Bank, External Affairs Unit, African Region, p. 2.

Yolove, E.A. "Nigeria: System of Education" in <u>The International Encyclopedia of Education</u>. 2nd ed. by Torsten Husen & P.N. Postlethwaite. (Pergamon), 1994.

Zahra, S.A. "What supervisors think about quality circles." <u>Supervisor Management</u>, 29, 1984.

About the Author

Dr. Frank Chika Okechukwu earned the Ph.D. in International Business from Clark-Atlanta University in Atlanta, Georgia in 1998. He earned the MBA in Finance from Morgan State University in 1991. His bachelor's degree in Management/Finance was earned in 1987 from the University of Sokoto, now Usmanu Dan Fodio University, in Nigeria.

He is a television and radio personality who has also authored many newspaper publications. He has taught International Business and Finance at the College of Notre Dame of Maryland; Business Law at Baltimore City Community College; and Marketing at Morgan State University. He is currently the Assistant Director of Bon Secours Baltimore Health System's Next Passage program.

These pages intentionally left blank.

These pages intentionally left blank.

www.ingramcontent.com/pod-product-compliance
Lightning Source LLC
Chambersburg PA
CBHW030359290526
45785CB00004B/1828